ELECTRICAL SYSTEM DESIGNING MADE SIMPLE

ELECTRICAL SYSTEM DESIGNING MADE SIMPLE

RAJIV SHANKAR

Viva Books

New Delhi | Mumbai | Chennai | Kolkata | Bengaluru | Hyderabad | Kochi | Guwahati

First Published 2007
Reprinted 2009, 2011

VIVA BOOKS PRIVATE LIMITED

4737/23 Ansari Road, Daryaganj, New Delhi 110 002
E-mail: vivadelhi@vivagroupindia.net, Tel. 42242200

76, Service Industries, Shirvane, Sector 1, Nerul, Navi Mumbai 400 706
E-mail: vivamumbai@vivagroupindia.net, Tel. 27721273, 27721274

Jamals Fazal Chambers, 26 Greams Road, Chennai 600 006
E-mail: vivachennai@vivagroupindia.net, Tel. 28290304, 28294241

B-103, Jindal Towers, 21/1A/3 Darga Road, Kolkata 700 017
E-mail: vivakolkata@vivagroupindia.net, Tel. 22836381, 22816713

7, Sovereign Park Apartments, 56-58, K. R. Road, Basavanagudi, Bengaluru 560 004
E-mail: vivabangalore@vivagroupindia.net, Tel. 26607409, 26607410

101-102 Mughal Marc Apartments, 3-4-637 to 641, Narayanguda, Hyderabad 500 029
E-mail: vivahyderabad@vivagroupindia.net, Tel. 27564481, 27564482

Beevi Towers, First Floor, SRM Road, Kaloor, Kochi 682 018, Kerala
E-mail: vivakochi@vivagroupindia.net, Tel: 0484-2403055, 2403056

232, GNB Road, Beside UCO Bank, Silpukhuri, Guwahati 781 003
E-mail: vivaguwahati@vivagroupindia.net, Tel: 0361-2666386

www.vivagroupindia.com

ISBN : 978-81-309-1733-7

Published by Vinod Vasishtha for Viva Books Private Limited,
4737/23, Ansari Road, New Delhi-110 002.

Printed and bound by Anand Sons, Delhi - 110092

Contents

Acknowledgements *ix*

Introduction *xi*

Chapter 1: General Aspects of Electrical Substation Designing **01-28**

1.0 Introduction 01
1.1 Different Types of Loads 02
1.2 Estimation of Load 05
1.3 Choice of Voltage 09
1.4 Safety, Reliability and Flexibility in Design 10
1.5 Space Requirement and Choice of Indoor/Outdoor Substation 11
1.6 Environmental Conditions 12
1.7 Area Classifications 13
1.8 Conclusion 22
1.9 Software 23

Chapter 2: Transformer **29-56**

2.0 Introduction 29
2.1 Selection of Transformer 29
2.2 Comparison between Dry-Type and Oil-Filled Transformer 32
2.3 Specifications 34
2.4 Installation 35
2.5 Safety Requirements 36
2.6 Commissioning 37
2.7 Maintenance 37
2.8 Other Important Aspects of Transformer 40
2.9 Evaluation of Transformer Losses 42
2.10 Conclusion 47
2.11 Software 48

Chapter 3: Power Control Centre

Chapter 3: Power Control Centre 57-104

3.0 Introduction 57
3.1 Bus-Bar 58
3.2 Switching Devices 61
  3.2.1 HT circuit breaker 61
  3.2.2 Comparison between different types of circuit breakers 63
  3.2.3 Specifications 67
  3.2.4 Low-voltage protective devices 67
3.3 General Requirement of Power Control Centre (PCC)/ Motor Control Centre Panel (MCC) 77
3.4 Protection and Measuring Schemes 78
3.5 Installation 91
3.6 Testing 92
3.7 Future Trends in Power Control Centre and Motor Control Centre 92
3.8 Conclusion 96
3.9 Software 97

Chapter 4: HT and LT Cables 105-124

4.0 Introduction 105
4.1 Construction 106
4.2 Specifications 108
4.3 Selection 108
4.4 Tests 109
4.5 Installation 109
4.6 Economics in Selection of Cable 111
4.7 Conclusion 112
4.8 Software 113

Chapter 5: Bus-Duct 125-136

5.0 Introduction 125
5.1 Selection 125
  5.1.1 Ambient Conditions 126
  5.1.2 Material 126
  5.1.3 Rating 127
  5.1.4 Enclosure 128
5.2 Calculation of Bus-Bar Size 129
  5.2.1 De-rating due to Temperature Rise 129
  5.2.2 De-rating due to Enclosure 129
  5.2.3 De-rating due to Altitude 130
5.3 Specifications 130
5.4 Conclusion 131
5.5 Software 132

Chapter 6: Capacitor Bank 137-148

6.0 Introduction 137
6.1 Reactive Power 137
6.2 Power Factor 138
6.3 Remedy for Poor Power Factor 138
6.4 Location of Capacitor Bank 139
6.5 Economics of Power Factor Improvement 140
6.6 Conclusion 145
6.7 Software 145

Chapter 7: Grounding 149-166

7.0 Introduction 149
7.1 Types of System Grounding 150
7.1.1 Solid grounding 151
7.1.2 Resistance grounding 153
7.1.3 Reactance grounding 153
7.1.4 Transformer grounding 154
7.2 System Grounding Points 155
7.3 Calculation of Ground Fault Current 156
7.4 Equipment Grounding 156
7.5 Design of Earth Electrodes 158
7.5.1 Plate electrode 159
7.5.2 Pipe electrode 159
7.5.3 Strip or conductor electrode 159
7.6 Conclusion 161
7.7 Software 161

Chapter 8: Battery Bank and Battery Charger 167-184

8.0 Introduction 167
8.1 Lead-Acid Battery 167
8.2 Nickel-Cadmium Battery 168
8.3 Selection 169
8.4 Environment 170
8.5 Comparison between Wet and Sealed Low-Maintenance Battery 170
8.6 Sizing of the battery 171
8.6.1 Rate of discharge 171
8.6.2 Temperature correction 172
8.6.3 Ageing factor 172
8.6.4 End cell voltage 173
8.7 Comparison between Lead-Acid and Nickel-Cadmium Battery 173
8.8 Memory Effect 175
8.9 Sulfation 175
8.10 Specifications of the Battery 176

8.11 Battery Charger 177
8.12 Specifications of Battery Charger 178
8.13 Tests 180
8.14 Capacity Test 180
8.15 Conclusion 180
8.16 Software 181

Chapter 9: Diesel Generating Set **185-198**

9.0 Introduction 185
9.1 Selection of DG Set 185
9.1.1 Non-linear load 185
9.1.2 Capacitive load 186
9.1.3 Unbalance load 186
9.1.4 Motor load 186
9.2 Assessing the Load 187
9.3 Location for Installation of DG Set 187
9.4 Specifications 187
9.5 Conclusion 188
9.6 Software 189

References and Bibliography **199**
Indian Standards **201**
Index **203**

Acknowledgements

I would like to express my special gratitude to Mr. K. Das, Ex-Executive Director of Krishak Bharati Cooperative Ltd, for reviewing the book and providing valuable suggestions.

I am thankful to the management of Krishak Bharati Cooperative Ltd. for giving me permission to publish this book.

I also wish to thank all my colleagues and my family for their support and encouragement while I was writing the book.

Finally, I am thankful to Viva Books for publishing the book.

Introduction

The book covers basically all the aspects of electrical substation designing. A software is supplied along with this book. The software is user friendly and can do all the basic calculations required to be performed at the time of designing a substation. This will help the user in performing various calculations just by clicking the mouse of the computer instead of searching for formulas and performing calculations on paper. Printouts of all the calculation can be taken. The software can also generate specifications of various electrical equipments. This will help the user in preparation of specifications which are normally required to be attached with tender documents at the time of preparation of tender. It also contains a library where details of various electrical equipments manufactured by various manufacturers can be stored, retrieved and deleted by the user. This will help the user to have a quick reference of the parameters of the equipments available in the market.

The book contains various numerical examples, and all the calculations given in the examples are also done by using the software to have a proper verification of the built-in formulas in the software.

A step-by-step procedure of performing various calculations, generation of specifications, storing, retrieving and deleting the information in the library is given at the end of each chapter.

This book will be useful for electrical consultants, practising engineers and students.

A brief description of each chapter is given below:

Chapter 1: General Aspects of Electrical Substation Designing

Chapter 1 provides the basic information required for designing of a substation. Various types of loads, estimation of load, aspects of safety, reliability and flexibility in substation designing and influence of ambient conditions on electrical equipments have been discussed and supported by examples.

Chapter 2: Transformer

This chapter deals mainly with the practical aspects of the transformer, like comparison between dry type and oil type of transformers, installation, testing and commissioning, specifications, safety requirements, oil testing, significance of percentage impedance, phenomena of inrush current at the time of switching of transformer.

Chapter 3: Power Control Centre

In this chapter different types of bus-bar schemes generally used in electrical system of an industrial plant, their advantages and disadvantages, specifications of HT circuit breaker, comparison of different types of HT circuit breakers, specifications of LT switchgear, comparison of air circuit breaker, MCCB, MCB, and fuses have been discussed. This chapter also contains aspects of protection and measuring schemes of power control centre, fault calculations like three phase fault, phase to phase, phase to phase and earth, phase to earth fault at HT end, LT end, and load end and relay coordination. Installation, testing and commissioning of power control centre and future trends in construction of power control centre have also been discussed.

Chapter 4: HT and LT Cable

This chapter contains details of construction of HT and LT cable, comparison between XLPE and PVC insulation, selection and economics of cable and voltage drop calculations.

Chapter 5: Bus-Duct

In this chapter, selection, construction and calculation for sizing of bus-duct have been discussed.

Chapter 6: Capacitor Bank

This chapter contains discussion regarding advantages of improved power factor, calculation of KVAR requirement to improve the existing power factor to target power factor and saving in electricity bill.

Chapter 7: Grounding

This chapter provides in-depth details of various types of system grounding, design of earth electrode, advantages of system grounding, and equipment grounding. It also provides details of different types of ground fault protection schemes.

Chapter 8: Battery

In this chapter discussion is confined to nickel-cadmium and lead-acid batteries because these two types of batteries are most commonly used in substations as a reliable/ emergency sources of power supply. The construction of various types of lead-acid batteries and pocket-plate nickel-cadmium batteries are discussed. A comparison between wet and sealed type batteries , nickel-cadmium and lead-acid batteries have been done. Various aspects related with battery like sulfation, memory effect etc. have been discussed. Specifications of battery charger and battery sizing calculation are also mentioned.

Chapter 9: Diesel Generating Set

In this chapter selection of DG Set, specifications, effects of various loads on performance of a DG Set, sizing of DG Set and voltage drop calculations are given.

What's on the CD-ROM

The CD-ROM consists of a software program for performing various calculations for designing an electrical system. The program is user friendly and performs the various calculations with the click of the mouse on the specified buttons. The facility for storage of data of various equipments is also provided in the software program. Besides this the specifications can also be generated by using the program. The software can easily be installed on the computer.

Installation of the software

1. The CD-ROM contains a folder "SSD".
2. Copy the "SSD" folder from the CD to C:\WINDOWS.
3. Open the "SSD" folder.
4. Click on "Edit" on the main menu bar and click on "Select All".
5. Click on "File" on the main menu bar and then click on "Properties".
6. Go to "Attributes" on the "Property" dialogue box.
7. Deselect "Read-only" and select "Archive" and click on "Apply" and then click on OK box. The "Property" dialogue box will close.
8. Open the Package folder and double click on "Setup.exe" icon.
9. Sub-station Designing screen will open up. Click on OK button.
10. Click on installation button of "Station-designing software".
11. After a little while a message box with the message "Sub-station designing set up was completed successfully" will appear.
12. Close the folder.
13. Go to the program and click on "Sub-station designing" to run the program.
14. It is necessary to copy the folder in C:\WINDOWS, otherwise the functions like "specifications" and "library" will not work.

1 General Aspects of Electrical Substation Designing

1.0 Introduction

Suppose, you have been appointed as an electrical consultant and assigned a job of designing and preparing of specifications of equipments of an electrical substation. Now, the question arises about how to proceed. The first step is to collect the following basic data:

a. Present load to be fed by the substation and future increase in load.

b. Voltage at which power will be received and the voltage at which load will be fed.

c. Type of load like lighting load, motor load etc.

d. Normal load and emergency load (load supplied from emergency source like DG Set, UPS etc.).

e. Space requirement and location of substation, like indoor or outdoor type.

f. Environmental conditions like maximum ambient temperature, humidity, height from sea level, clean or dusty environment etc.

g. Whether the substation is going to be located in hazardous area and flame-proof equipments are required.

h. Spare capacity of substation equipments.

i. Scope of further addition of spare feeders.

After fixing the boundaries, we should go into more details like fault level of equipments, details of incoming and outgoing feeders, changeover scheme, redundancy of equipments, protection scheme, electrical parameters measuring and recording schemes, bus-duct, cable, cable tray arrangement, type of insulation for high and low tension equipments, safety requirements, and of course the cost of the substation.

In this chapter we will discuss the general aspects of designing a substation. A case study in the end of this chapter will also be taken up. Besides this, we will also see how the software which is supplied along with this book works for calculation of load.

1.1 Different Types of Loads

Before designing a substation, it is a foremost requirement that the present substation load and the future demand be correctly accessed. The present load and the estimation of future demand influence the sizing of the individual equipment. The load depends on the nature of area like residential, industrial, commercial, rural etc.

Now, as a first step in estimation of load, a detailed load schedule shall be prepared. Before preparation of load schedules, a detailed interaction is needed with the consultant/architect, equipment manufacturer, process engineers as the case may be, to have a correct idea of the continuous running load and emergency load. Different types of load should be segregated. In load schedules, various loads are put in a tabular form and the total load is calculated by adding them. Various factors like demand factor, diversity factor, load factor are applied to have a fair idea of the maximum demand of the system. These factors are explained in the following sections. Equipment should be capable of taking care of maximum calculated demand. In addition to that, some spare capacity of the equipment is also kept for future expansion. The practice of keeping hundred per cent spare capacity is more prevalent in process industry where availability of system is more critical, and breakdown in electrical system may hamper the entire process and loss of production will take place to a great extent.

Past data, graphs and charts of similar type of establishment are also helpful in estimation of load. For example, the graph in the Fig. 1.1 gives us an idea of variation of load of a process plant.

While estimating the load, the duty cycle of each equipment shall also be taken into account. For example, in a workshop for a particular operation, a group of motors are provided, but all the motors may not be working simultaneously.

In the process industry, it is a usual practice to provide oversize motors to increase the reliability. For example, vendors are asked to provide motors with class 'F' insulation, but the temperature rise should be limited to class 'B' insulation. The class 'F' insulation can withstand upto 155^{0}C, while class 'B' can withstand upto 130^{0}C. This means that vendor will provide the motor one size higher than the required size, resulting in slightly oversizing of motor as also the cost. Therefore, a trade-off is needed between the cost and reliability.

The substation should be capable of meeting the peak demand. It is always preferred to have a flat demand curve. If the demand curve is not flat and the substation

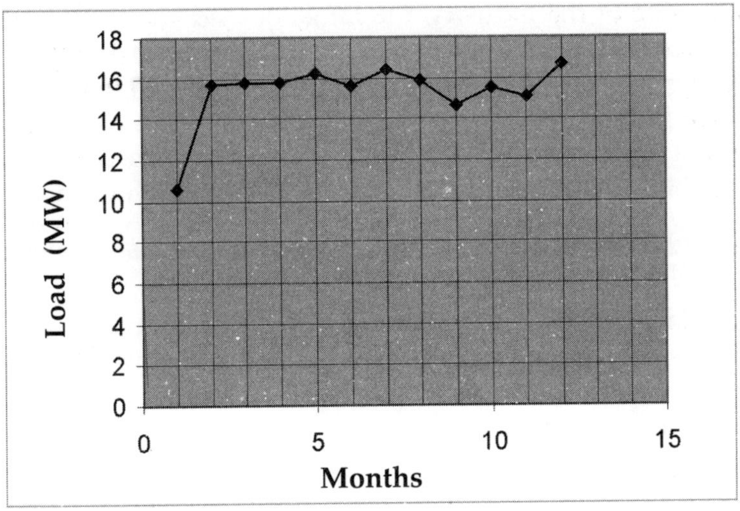

Figure 1.1: Variation of load of a process plant

experiences peak for a short duration, then it is preferred to supply this short peak demand through a stand-by source. For example, a diesel generating set can be used to cater for the peak demand, which is for a short duration instead of augmentation in the capacity of substation.

Emergency loads are those critical loads for which power supply is essentially required when the normal power supply fails. For example, in a process industry to have a safe shut down and to keep the flow of process fluids, certain critical pumps are required to be started immediately through alternate power source in case of failure of normal power supply. The emergency loads are connected to additional source of power supply like diesel generators, UPS etc., and a changeover from normal source to emergency source takes place in case of failure of normal power supply. The other emergency loads on which safety of humans depends, like fire alarm system and fire fighting system, lifts, lighting etc., should be connected to an alternate power source in case of failure of normal power supply.

In Example 1.1, the emergency and normal load of a commercial and industrial complex are mentioned.

Broadly, the load can be divided into the following categories:

- Industrial
- Commercial
- Domestic

The industrial load may be categorized as per the size of the industry, like cottage industry, small-scale industry, medium-scale industry and large industry.

According to the size of the industry, the range of load varies. For example, a large industry consumes bulk load in the range of few thousand kilowatts. Bulk load is supplied at higher voltage like 11 KV, 33 KV and 66 KV. The load requirement for small and cottage industry may be less and in the range of 5 KW to 100 KW and the load is supplied at low voltage i.e. 415 V. For the medium size industry, the load may be around a few hundred KW and normally supplied at high voltage i.e. 11 KV.

It is always economical to supply bulk load at high voltage to reduce the size of conductor and transmission losses.

Load of commercial establishment consists of lights, fans, lifts, airconditioners, display boards, computers, UPS, motors for firefighting and drinking-water pumps etc. The load may vary from a few kilowatts to a few hundred kilowatts, depending on the size of commercial establishment.

Domestic load consists of lights, fans, air-conditioners, heaters, lifts (high rise buildings).

In a large industry, for different load centres, separate electrical substations are required.

Example 1.1

Example of different types of load

Table 1.1
Commercial Load

Normal Load	Emergency Load
Area lighting	Fire alarm system
Hall lighting	Firefighting system
HVAC system	Lifts
Cabin lighting	UPS and computers
Fans	Drinking water system
	Air handling units
	Lighting
	Staircase
	Lift lobby
	Corridor
	Toilet, pantry
	Electrical substation
	Car parking
	AC plant

Table 1.2
Industrial Load

Normal Load	Emergency Load
Area lighting Workshop Testing devices Heating equipment Electrical drives for process/ production equipment	Hoist Safety equipment Firefighting system Fire alarm system UPS, computers, lighting and any critical process equipment

Normal and emergency loads can be of continuous or intermittent type.

1.2 Estimation of Load

Various factors are used in load estimation:

- Demand factor
- Load factor
- Diversity factor

Demand factor

The demand factor is the ratio of maximum demand and the total connected load. It is either 1 or less than 1. If a consumer does not use all the connected load at a time, the demand factor is less than 1. Normally, in case of domestic and commercial load, the demand factor is less than 1. This can be further explained with the help of Example 1.2.

Example 1.2

Suppose a domestic load consists of 10 fluorescent lamps of 55 watt, 5 fans of 70 watt. 6 lamps and 3 fans are used, the maximum demand shall be (6 x 55 + 3 x 70) = 540 watt.
The connected load will be (10 x 55 + 5 x 70) = 900 watt.
Demand factor: 540/900 = 0.6

Load factor

The load factor is the ratio of average load to maximum load. The load factor can also be defined as the ratio of energy consumed during a period of time to the energy that would have been consumed if the demand had been maintained maximum during that period. Refer to Example 1.3.

Example 1.3

Suppose a domestic consumer has the following load:

550 watt from 6 pm to 10 pm
350 watt from 10 pm to 6 pm
The energy consumed: (550 x 4 + 350 x 20) = 9200 watt-hrs
Maximum energy: 900 x 24 = 21600 watt-hrs
Load factor: 9200/21600 = 0.426
Normally the load factor is less than 1.

Diversity factor

Diversity factor is the ratio of simultaneous maximum demand and the sum of individual maximum demand.

Example 1.4

Suppose, there are four consumers and the maximum load for a particular time is as follows:

Consumer 1: 400 watt
Consumer 2: 300 watt
Consumer 3: 250 watt
Consumer 4: 200 watt

Sum of individual maximum load: 1150 watt
Simultaneous maximum demand: 4 x 400 = 1600 watt
Diversity factor: 1600/1150 = 1.39
Now the question arises as to what should be the load factor and diversity factor. This depends on one's experience. For assessment of load, some of the state electricity boards have fixed the load factor and diversity factors for different types of consumers. Refer to Table 1.3, which gives an idea of load factor and diversity factor for different types of consumers.

Table 1.3

Sl. No.	Category	Purpose	Diversity factor	Load factor
1.	Residential	Lighting and fans	2.5	30%
		Heating appliances	1.0	15%
		Cooling appliances		
		Feb-Oct	1.0	50%
		Nov-Jan	1.0	NIL
2.	Non-residential	Lighting and fans	1.0	35%
		Heating appliances	1.0	50%
		Cooling appliances	1.0	40%
3.	LT (Industrial)	Engineering Workshop	1.5	30% for one shift, 50% for two shifts, 70% for three shifts. (Each shift of eight hours)
		Power looms and textile	1.2	30% for one shift, 55% for two shifts. (Each Shift of 12 hours)
		Ice-cream manufacturing		
		Feb-Oct	1.0	80%
		Nov-Jan	1.0	80%
		Ice Factory		
		Feb-Oct	1.0	80%
		Nov-Jan	1.2	40%
		Cold Storage		
		Feb-Oct	1.0	75%
		Nov-Jan	1.0	50%
		Society Water Pumps	1.0	25%
		Cinema/theatres	1.5	60%
		Flour Mills	1.0	30%
		Ginning and Pressing	1.2	60%
		Foundry	1.5	50%
		Plastic	1.2	50%
		Rubber	1.2	50%

	Mining, Quarry and stone crushing	1.2	60%
	Chemical industry Textile units	1.2 1.2	60% 70%
	Others	1.5	30% for one shift, 60% for two shifts, 80% for three shifts. (Each shift of eight hours)

HT Consumers

Category of HT Consumers	Load Factor
Hotels	65%
Iron and Steel	60%
Foundry	60%
Steel rolling mills	60%
Chemical factory	80%
Paper mills	85%
Textile	75%
Cement	70%
Fertilizers	70%
Oil mills	70%
Solvents	70%
Ginning and pressing	60%
Engineering goods	50%
Ice	75%
Cold storage	75%
Sugar and Khandsari	60%
Dairy	75%
Mining, quarry and stone crushing	50%
Plastic and rubber	60%
Cinema	60%
Others	
8 hrs shift	30% – 1 shift
	60% – 2 shifts
	90% – 3 shifts
	12 hrs shift
	45% – 1 shift
	90% – 2 shifts

1.3 Choice of Voltage

Load is supplied at different voltage levels. For distribution system the voltage of the feeding line depends on the load. It is always economical to receive supply at higher voltage if the demand is more. If a large load is fed at low voltage, the size of the transmission line, transformer etc. will increase so that it can supply larger amount of current. The transmission losses will also increase.

Normally, the electricity boards have their own norms for fixing the supply voltage for feeding loads of different magnitudes. For example, load less than 100 KW is supplied at 415 V, load between 100 KW and 3000 KW is supplied at 11 KV and if the load is above 3000 KW, it is supplied at 33 KV. Further, it also depends on whether the transmission line of required voltage is available in the vicinity of load and if it is available, then again the question arises whether it has the capacity of delivering the required load or not. If the load is supplied at a higher voltage level, the voltage regulation will be good and the load losses of transmission line will be much less. Refer to Example 1.5 to have an idea of voltage levels and limits of variation of voltage in single phase and three phase electrical system.

Example 1.5

Voltage levels

Standard Voltages:

Single phase two wire system: The standard voltage for single phase two wire system, is 240 V.

Three phase system: The standard voltages for three phase system are as follows:

415 V, 3.3 KV, 6.6 KV, 11 KV, 22 KV, 33 KV, 66 KV

110 KV, 132 KV, 220 KV, 400 KV

Voltage limits for AC system

Under normal conditions, the variation in the voltage shall not be more than the limit mentioned below:

- 6% in case of low or medium voltage
- 6% on the higher and 9% on the lower side in case of high voltage
- 10% in the higher side and 12.5% in the lower side in case of extra high voltage

As per the IE rules, the categorization of voltages are as under:

Low voltage: The voltage which does not exceed 250 V is called low voltage.

Medium voltage: The voltage which exceeds 250 V but does not exceed 650 V is called medium voltage.

High voltage: The voltage which normally exceeds 650 V but does not exceed 33 KV is called high voltage.

Extra high voltage: The voltage exceeding 33 KV under normal condition is extra high voltage.

1.4 Safety, Reliability and Flexibility in Design

Safety

Safety is the most important factor in the design of the electrical system. The established codes should be followed, which takes care of the safety aspects. Some of the safety points are mentioned below:

- Circuit breakers, MCCB or any other switching device should be able to function properly under the most severe duty for which they are expected to expose.
- No live part should be exposed and protection must be provided against accidental contact with the live part of the electrical system. The bus-bars should be properly sleeved and joints should be shrouded.
- Proper grounding of each and every equipment should be provided.
- Proper interlocking between upstream and downstream breakers should be provided to prevent any reverse power feedback.
- Protective relays should be properly coordinated.
- Oil soak pits should be provided, if oil type transformers are installed.
- Cable entry and exit points in substation should be properly sealed with a fire-resistant compound to prevent propagation of fire.
- Segregation of transformers with two hours fire rating wall should be done, if more than one transformer is provided.

Reliability

It depends on the type of application. Some types of industrial load can tolerate a little interruption, while others, particularly the process industry, cannot tolerate the interruption and require electrical system of high reliability. In such a system, redundancy is kept. Provision for emergency power supply from another source like diesel generating sets, UPS etc. is also kept. Further, the load which is very sensitive to variation in supply voltage and frequency like computers, programmable logic controllers and DCS system should be supplied through UPS system.

Flexibility

The system should be flexible enough so that it can accommodate the increased future demand which may arise due to expansion/addition in the existing plant/machinery etc. during the life of the plant.

What should be the provision for future growth? This is difficult to estimate, but one should have a rough idea about the growth of demand. The demand of a residential/ commercial area increases with the growth of population, rise in living standard of the people, development of infrastructure facilities etc. Similarly, in the industrial sector, expansion/diversification takes place and accordingly, the demand increases. If more than one substations are provided and each has spare capacity, then these substations can pool their spare capacity to meet the additional demand.

Power factor

The power factor of a substation should never be less than 0.95. If it is low, capacitor banks should be provided to improve the power factor. This aspect has been discussed in more detail in Chapter 6.

1.5 Space Requirement and Choice of Indoor/Outdoor Substation

In big metropolitan cities, space is costly, and it is a normal practice to use minimum space for the services like electrical substation, central air-conditioning plant and fire pumps. If, the substation is being set up for feeding a commercial load or residential load, then the obvious choice is indoor type of substation. The reason is that the indoor type of substation occupies less space and aesthetically too looks suitable. However, in case of a large industry where load is fed at high or extra high voltage level, one has no choice except to set up an outdoor switch yard. There may be more than one substation in a large industry, because it is economical to set up a substation near the load centre.

The requirement of space depends on the size of the equipment. A proper layout should be developed and the following points shall be taken care of.

- Sufficient space for operation and maintenance should be provided. Refer to Chapter 3/5.5 for the distances.
- Sufficient clearance should be maintained between switchgear and room walls.
- Exit doors and wall openings of sufficient sizes should be provided for removal and entry of electrical switchgear.
- Ventilation of room should be proper so that the electrical equipments can perform as per the temperature limit provided for them. Otherwise if the temperature rises beyond the specified limit, de-rating of equipment will take place.

- Cable trenches of proper size should be provided, and it is preferred that the trenches be filled with sand. Alternatively, overhead racks may be constructed for laying cables.

A typical layout of indoor substation is given in Fig. 1.2. This layout is for two transformers of 750 KVA capacity.

It is preferable to have LT, HT panels, battery chargers etc. on the first floor and cable alley on the ground floor. With this arrangement, underground trenches can be avoided. It is also preferable to have bus-duct at the LT side of the transformer though the cost of bus-duct will be more than that of the cables. The reason is that the numbers of cables are more at LT side and therefore space requirement will be more. Number of terminations will also be more resulting in more chances of failure of termination. The current may not be equal in all parallel cable due to variation in cable impedance. Further, if the length of cable is more, voltage drop will also be more.

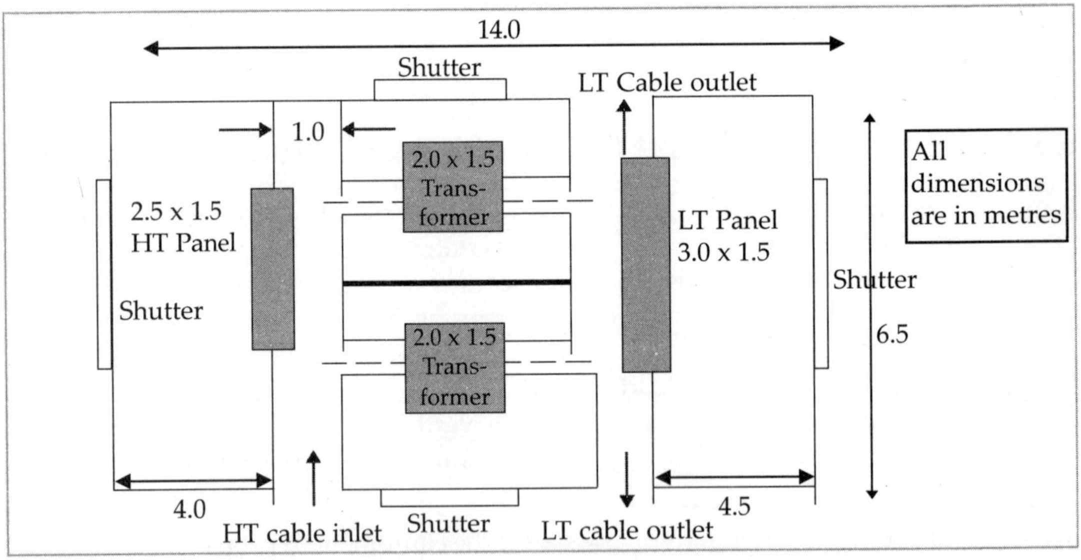

Figure 1.2: Typical substation layout

1.6 Environmental Conditions

Performance of the electrical equipments depends on the environmental conditions. The following factors are taken into account while making selection of the electrical equipments.

- Ambient temperature
- Altitude
- Relative humidity

The most important is the ambient temperature since the equipments are designed to perform under a certain temperature range, and beyond that limit (higher limit) equipments are to be de-rated. Normally, maximum and minimum ambient temperatures are specified with the equipment specifications.

Most often the reference ambient temperature and the maximum ambient temperature are considered the same, although they are not. As per IS 9676–1987, the reference ambient temperature is 1/3 of $(Qp+Qd+Qy+10)^0C$

Qp = Maximum ambient air temperature
Qd = Maximum daily average air temperature
Qy = Maximum yearly average temperature

As per Indian standard, 40^0C is the standard reference ambient temperature for electrical equipments.

It is possible to calculate the reference ambient conditions and specify it in the specifications. Maximum ambient temperature is the maximum temperature that may have reached at a particular place for a short period of time. The temperature rise is very important since it is directly proportional to the square of the current and a small difference in temperature rise affects the size of the equipments considerably.

The altitude also affects the sizing of the equipments, since as the altitude increases, the air becomes thin and it lowers the cooling of the equipments during the running conditions. Normally, equipments are designed for an altitude of 1000 metres. For a higher altitude de-rating of equipment is required.

1.7 Area Classifications

In some industries like petrochemical, fertilizers etc. classification of various zones within the project areas should be specified as hazardous or non-hazardous according to the requirement of IS/IEC/NEC. The equipment shall be used accordingly. For example, explosion-proof equipments having protection such as EX-d or Ex-e are used in accordance with hazardous area classification in the chemical plant.
Now, we take a case study for calculation of load in a multi-storeyed housing complex. Refer to Example 1.6.

Example 1.6

Load calculation for a multi-storeyed housing complex

The housing complex consists of three blocks. Block-1, Block-2 and Block-3. The number of dwelling units, area, type and electrical points are shown in Table 1.4.

Table 1.4

Sl. No.	Description	No. of houses	Area (sq. metre)	5A socket (Nos)	15A socket (Nos)	40W lamp (Nos)	Ceiling fan (Nos)	100W lamp (Nos)	Exhaust fan (Nos)
1.	**BLOCK-1**								
	Type-1	14	80	12	5	7	5	4	1
	Type-2	14	100	15	7	10	7	5	2
2.	**BLOCK-2**								
	Type-3	18	125	18	9	11	9	5	3
	Type-4	18	150	20	11	12	10	6	3
3.	**BLOCK-3**								
	Type-5	12	175	22	13	15	12	8	3

In addition to the above, the housing complex has got two number of lifts in each block, area lighting, fire and drinking water pumps. The details are given below:

Table 1.5

Sl.No.	Description	Qty (Nos)	Rating (KW)
1.	Lifts (8 Passengers)	6	5
2.	Fire pumps	1	20
3.	Drinking water pumps	2	10
4.	Common area lighting	45	0.055
5.	Street lighting	25	0.15

The housing complex also has a commercial complex. The load details are given in Table 1.6.

Table 1.6

Sl. No.	Description	Nos.	Area (sq. metre)	5 A socket (Nos)	15 A socket (Nos)	40 W lamp (Nos)	Ceiling fan (Nos)	100 W lamp (Nos)	Exhaust fan (Nos)
1.	Super market	1	125	20	15	25	10	15	1
2.	Office rooms	5	15	12	5	20	5	5	0
3.	Shops	5	10	15	10	10	10	5	0
4.	Restaurant	1	100	15	12	20	12	10	4

It is required to calculate the load for the residential as well as commercial set-up. There are two ways to accomplish the above task. One is to calculate the load as per the number of points provided, and the other is calculation of load based on covered area. Some electricity boards follow the covered area approach, while others go by the number of points provided in a dwelling unit. Number of points vary from one establishment to another. The above example is considered to illustrate the load calculation method. Now, we proceed step by step.

Step 1

Calculation of load requirements:

In Step 1 we take the covered area approach in calculating the load. Table 1.7 indicates the load based on covered area for domestic and commercial establishments. The figures mentioned in the table is just for illustration of the load calculation method based on covered area.

Table 1.7

Sl. No.	Description of load	Load as per approved FAR (floor area ratio)
1.	Commercial without air-conditioning system	15 KW/100 sq. mts.
2.	Commercial with air-conditioning system	25 KW/100 sq. mts.
3.	Domestic	12 KW/100 sq. mts.

Now the job has become pretty simple, and the connected load for each dwelling unit and also the total load of all the dwelling units have been calculated as per Table 1.8.

Table 1.8

Calculation of load of a residential complex based on the covered area

Sl. No.	Description	No. of dwelling units (A)	Flat Area (sq. metre) (B)	Load/Flat (KW)(BX 0.12) (C)	Total Load (KW) (CXA)
1.	**BLOCK-1**				
	Type-1	14	80	9.6	134
	Type-2	14	100	12	168
2	**BLOCK-2**				
	Type-3	18	125	15	270
	Type-4	18	150	18	324
3.	**BLOCK-3**				
	Type-5	12	175	21	252
	Total				**1148**

Table 1.9

Calculation of load of commercial complex on covered area basis

Sl.No.	Description	Nos. (A)	Area (sq. mtrs.) (B)	Load/unit (KW) (B X 0.25) (C)	Total Load (KW) (CXA)
1.	Supermarket	1	125	31.25	31.25
2.	Office Rooms	5	15	3.75	18.75
3.	Shops	5	10	2.5	12.5
4.	Restaurant	1	100	25	25
	Total				**87.5**

As calculated above, 1148 KW is the total connected load of residential complex and 87.5 KW is the total connected load of commercial complex. However, the substation will not be feeding this load continuously; otherwise one will end up in establishing a substation of very large capacity with very poor capacity utilization.

Therefore, it is required to apply demand factor and diversity factor to calculate the maximum demand on the substation. Now the question is, what should be the demand factor and diversity factor? There is no fixed criterion and as mentioned earlier, it is chosen depending on the individual's judgment.

Taking the demand factor as 0.7, the maximum demand will be 0.7 x 1148 = 803.6 KW for residential complex and 0.7 x 87.5 = 61.25 KW for commercial complex. Now this demand has diversity also. The residential and commercial loads are highly diverse loads and diversity factor upto 2 can be chosen. Applying diversity factor of 1.4 and 1.6 for residential complex and commercial complex respectively, the maximum demand for residential complex will be 803.6/1.4 = 574 KW and the maximum demand for commercial complex will be 61.25/1.6 = 38.28KW. The substation will experience the (574 + 38.28) KW = 612.28 KW, total demand of residential and commercial complex.

The load of common services is calculated as per Table 1.10.

Table 1.10

Common services

Sl. No.	Description	Qty (Nos)	Rating (KW)	Total Load (KW)
1.	Lifts (8 passengers)	6	5	30
2.	Fire pumps	1	20	20
3.	Drinking water pump	2	10	20
4.	Common area lighting	45	0.055	2.5
5.	Street lighting	25	0.15	3.75
	Total			**76.25**

Considering the demand factor 0.6 and diversity factor 1.5,

Maximum demand =76.25x0.6/1.5 = 30.5KW

Total Load = 612.28 + 30.5 = 642.78 KW say 643 KW. Considering the power factor 0.95 (assuming capacitor banks are provided), The maximum demand in KVA will be 643/ 0.95 = 676.84 KVA or say 677 KVA.

Since this is a housing complex and future expansion is not expected on a large scale, therefore considering a future increase in load of 10 per cent, the substation is designed for 677 x1.1 = 744.7 KVA demand.

Step 2

Calculation of emergency load

The emergency load of various types of dwelling units are taken as shown in table 1.11

Table 1.11

Sl.No.	Description	No. of dwelling units	Load/Flat (KW)	Total Load (KW)
1.	**BLOCK-1**			
	Type-1	14	0.5	7.0
	Type-2	14	0.6	8.4
2.	**BLOCK-2**			
	Type-3	18	0.7	12.6
	Type-4	18	0.8	14.4
3.	**BLOCK-3**			
	Type-5	12	1.0	12.0
	Total			**54.4**

Taking demand factor = 0.7 and

Diversity factor = 1.4

Maximum demand = 54.4 x0.7/1.4 = 27.2 KW

Table 1.12

Emergency load for comm‿n services

Sl. No.	Description	Qty (Nos)	Rating (KW)	Total Load (KW)
1.	Lifts (8 passengers)	3	5	15
2.	Fire pumps	1	20	20
3.	Drinking water pump	1	10	10
4.	Common area lighting	45	0.055	2.5
5.	Street lighting	25	0.15	3.75
	Total			**51.25**

Taking demand factor = 0.6 and

Diversity factor = 1.5

Maximum demand = 51.25 x 0.6/1.5 = 20.5 KW

Total Emergency Load = 27.2 KW + 20.5 KW = 47.7 KW, (59.63 KVA, assuming 0.8 power factor of the load , since capacitor banks are not recommended to operate when power supply is fed through D G Set)

Considering 10% future increase in demand, total maximum demand will be 1.1 x 47.7 = 52.47 KW. The alternative power supply source like DG Set should be chosen in such a way that it should be capable of supplying a maximum load of 52.47 KW (65.59 KVA at 0.8 power factor).

The load calculation has been summarized in table 1.13.

Table 1.13

Load Summary

Maximum demand for residential and commercial complex	643 KW (677 KVA)
Designed substation load (Considering 10% increase in load in future)	707.4 KW (744.7 KVA)
Emergency load for common services and residential complex	47.7 KW (59.63 KVA)
Designed substation emergency load (Considering 10% increase in load in future)	52.47 KW (65.59 KVA)

The maximum demand on substation will be 744.7 KVA with a 10 % margin for future growth. It is common practice to supply this load at 11 KV. Therefore a 11 KV substation is required to be set up. Now the question arises: what should be the capacity of the distribution transformers and will one transformer be sufficient for the entire load or are more transformers required? Technically, one transformer of 1000 KVA or even 750 KVA is sufficient for 744.7 KVA load. But what about the reliability/flexibility of the system? Will the system be reliable enough with one transformer? How much time will it take to repair the transformer if any fault takes place in the transformer, and during that period what will be the alternate source of power supply?

The answer to these above questions lies in the design philosophy, i.e. how much spare capacity is required and the expenditure needed to have that capacity.

It is assumed that 100% redundancy is required, i.e. two transformers are provided, and if one fails then the other transformer should be capable to take the entire load. Considering this, two transformers of 750 KVA capacity are chosen. A single line diagram of electrical system is shown in Fig. 1.3.

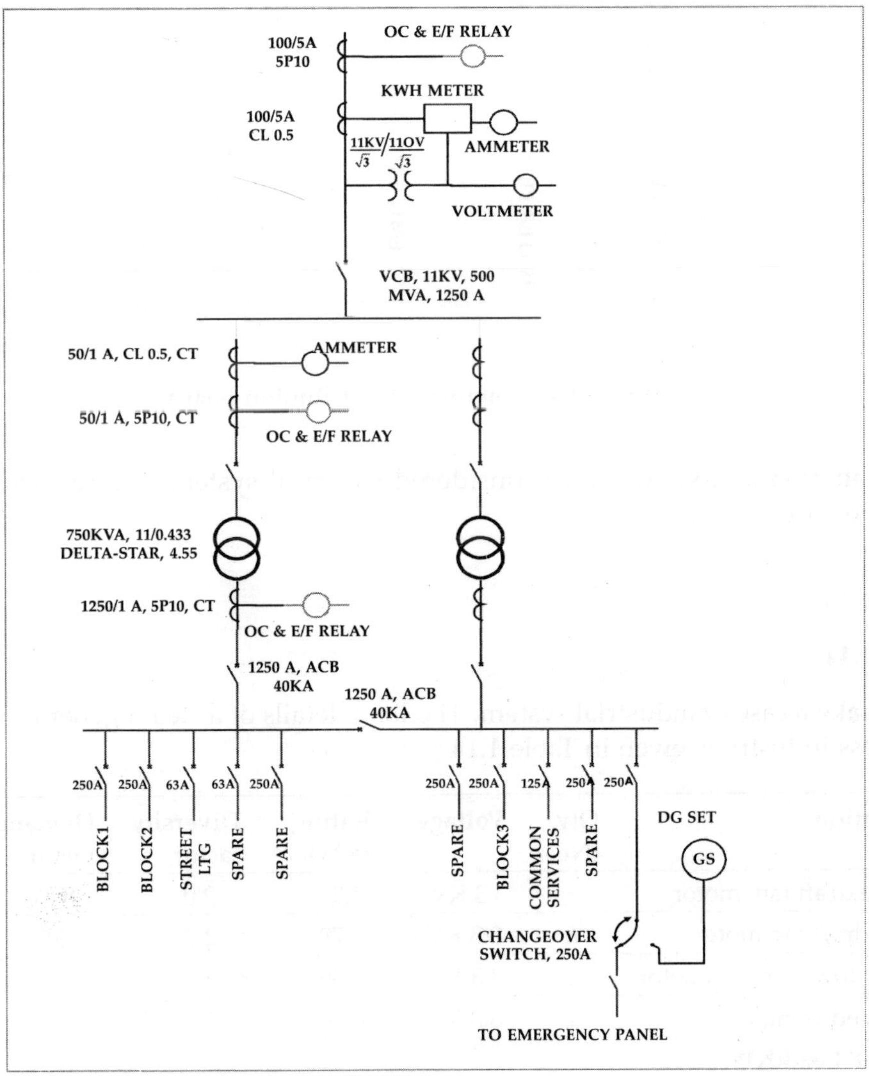

Figure 1.3: Single line diagram of electrical system

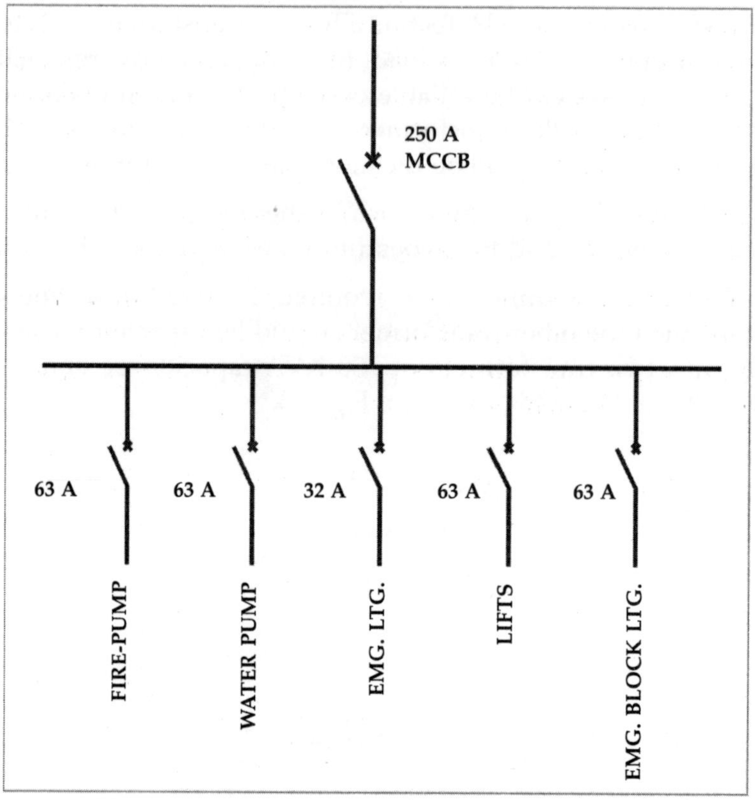

Figure 1.4: Emergency distribution board

In the above example, we have considered electrical system of a residential and a commercial complex.

Example 1.7

Table 1.14

Let us take a case of industrial system. The load details of a steam generation plant in a process industry is given in Table 1.14.

Description	Qty. (Nos)	Voltage	Rating (KW)	Diversity factor	Demand factor	Total (KW)
Induced draft fan motor	6	3.3 KV	430	2.0	80%	1032
Forced draft fan motor	6	3.3 KV	270	2.0	80%	648
Cooling tower pump motor	4	3.3 KV	335	2.0	80%	536
Boiler feed pump motor	2	3.3 KV	1800	2.0	80%	1440
Total HT Load(KW)						**3656**

Description	Qty. (Nos)	Voltage	Rating (KW)	Diversity factor	Demand factor	Total (KW)
Makeup water pump motor	3	415 V	90	1.5	80%	144
Air compressor motor	3	415 V	115	1.5	70%	161
Air heater	2	415 V	30	1	90%	54
Auxiliary water pump	2	415 V	75	1.5	80%	80
CT fan motor	4	415 V	45	1	60%	108
Condensate extraction pump	4	415 V	22	2	80%	35.2
Chiller compressor	3	415 V	55	1.5	80%	88
Air handling unit	3	415 V	45	1.5	80%	72
Auxiliary oil pump	2	415 V	22	2	70%	15.4
Air handling, cooling tower, and miscellaneous other load	6	415 V	17	2	80%	40.8
NGL firing pump	6	415 V	55	3	70%	77
NGL unloading pump	2	415 V	7.5	1.5	50%	5
Lighting, UPS and miscellaneous others	-	415V	15	1.8	80%	6.67
Total LT Load (KW)						**887.07**

The power for steam generation plant is received at 11 KV. It is to be stepped down at 3.3 KV and 0.433 KV. Since the electrical power supply should be more reliable in a process industry, a policy of 100% redundant power supply is adopted. Two step-down transformers will be provided with 100% redundant capacity so that if fault takes place in one of the transformer, the other transformer could take the full load.

Since the maximum demand for 3.3 KV motors is 3656 KW, two transformers of 4 MVA, 11/3.3 KV will be sufficient. Similarly, two transformers of 1000 KVA for 415 V system are chosen for maximum demand of 887.07 or say 887 KW.

A typical single line diagram for 3.3 KV system has been shown in Fig. 1.5.

A more detailed discussion will be done in the following chapters.

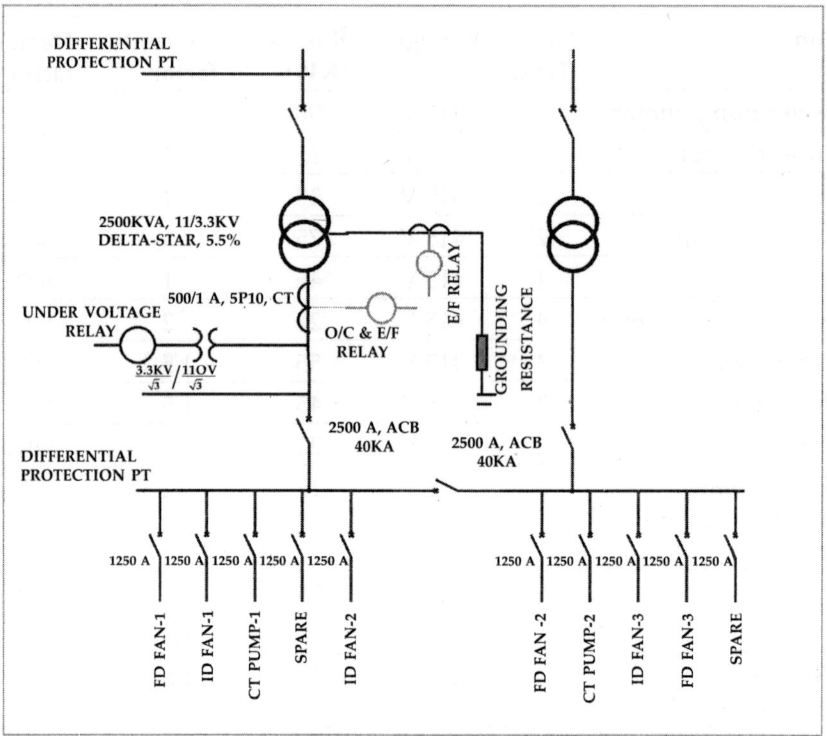

Figure 1.5: Single line diagram for 3.3 KV system of steam generation plant

1.8 Conclusion

In this chapter an attempt has been made to show how the estimation of load is done for a residential, commercial as well as industrial complexes. The calculations for normal and emergency loads have been shown in Tables 1.8 to 1.12.

After finalization of the maximum demand, the single line diagram has been developed. Before development of the single line diagram, other factors such as reliability, flexibility etc. are taken into account, and after that rating, capacity and sizing of equipment, type of distribution system such as radial, ring etc. are finalized. While choosing the system, in addition to the capital cost, life cycle cost of the equipment shall also be taken care of. The short circuit capacity is another factor, which should be carefully considered for each equipment. We will discuss all these aspects in subsequent chapters.

1.9  Software

Objective

- To calculate maximum demand of residential, commercial and industrial load
- To calculate maximum demand of common services like lifts, drinking water pumps, firefighting system etc.
- To calculate maximum demand based on covered area

How to use the software

Separate screens are provided for calculating different types of loads like housing, commercial and industrial. Load can be calculated in both ways, i.e. by considering the covered area as shown in the above example and also by considering the number of points in the dwelling units.

After installation of software, just start the program and click on the load calculation on the main menu bar Fig. 1.6. Fill in the project name.

Follow these steps.

- Fill in the Quantity, i.e. Number of points like light points, power points etc. for each type of dwelling unit in the specified boxes as per Table 1.15 in Example 1.8.
- Click the "Calculate" button to calculate the total connected load.
- Fill in the demand factor, diversity factor as per Example 1.8 and click "Calculate" button to calculate the maximum demand of the system.
- Above three steps are repeated for each type of dwelling unit.
- Click "Total Demand" button to add the loads of all types of dwelling unit.
- Click on common services to go to common services screen and fill in the required load as per the Table 1.10 in Example 1.6 in the respective boxes. Click the "Calculate" button to calculate the total connected load. Fill in the demand and diversity factors as given in Example 1.4 and click "Calculate" button to calculate the maximum demand of common services.
- To calculate the total load i.e. common services and the housing or commercial or industrial as the case may be, click on the respective tab and then click on "Total demand".
- On common services screen itself a table has been given to calculate the load by covered area method.

All the screens can be printed by clicking on the "Print" button.

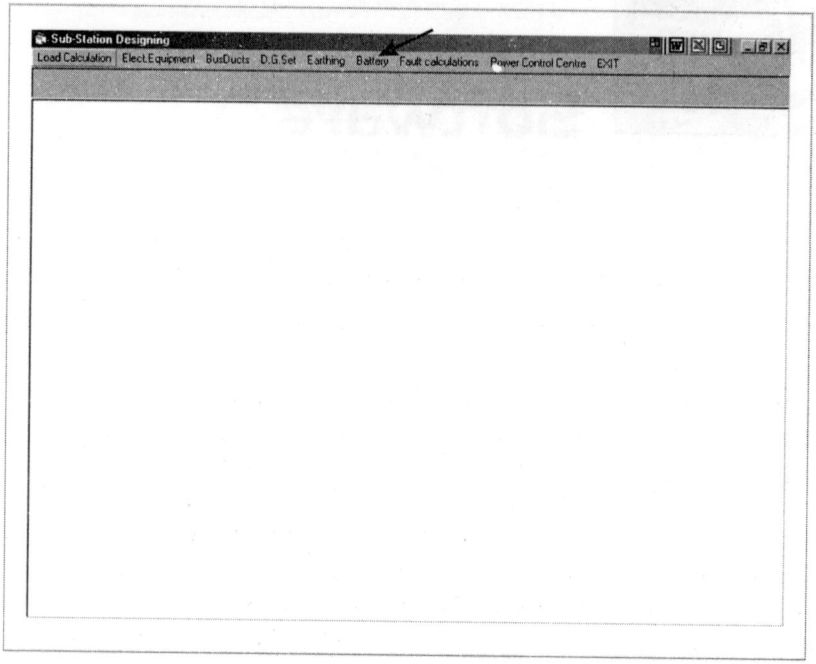

Figure 1.6: Main menu screen

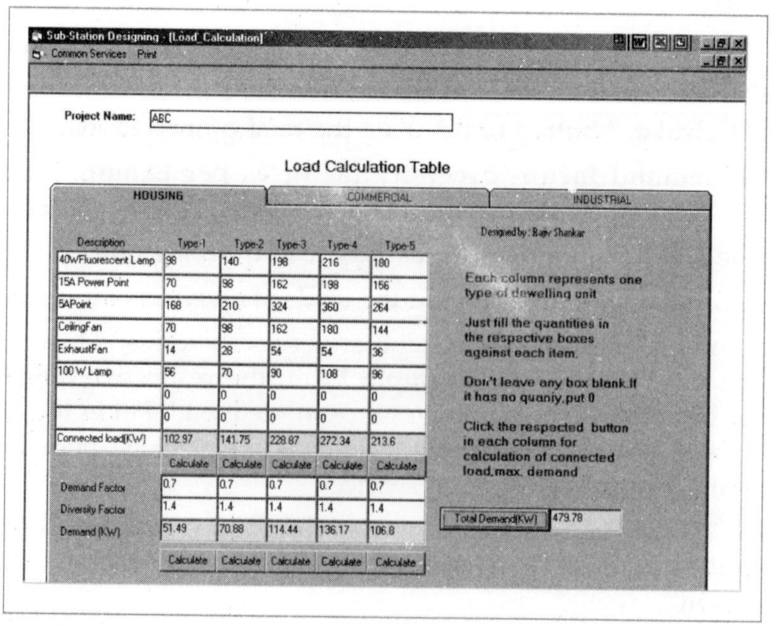

Figure 1.7: Load calculation screen of a housing complex

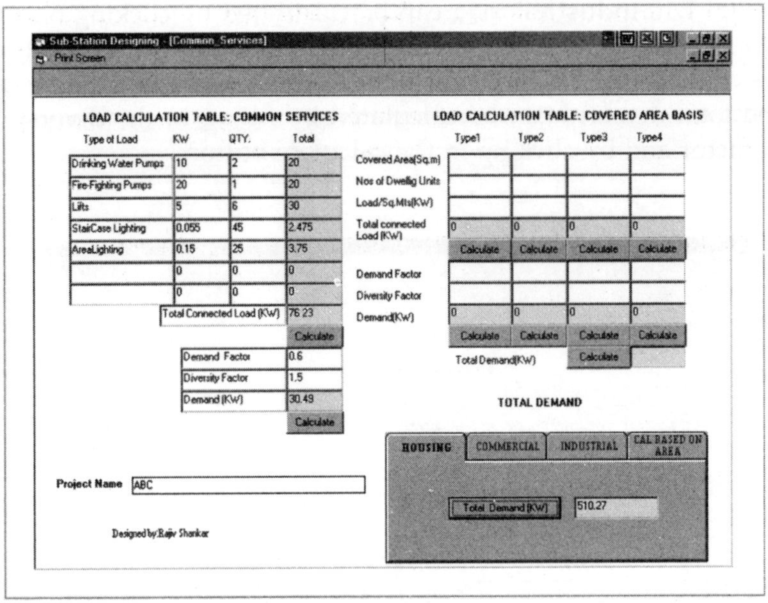

Figure 1.8: Load calculation screen for common services and load calculation on covered area basis

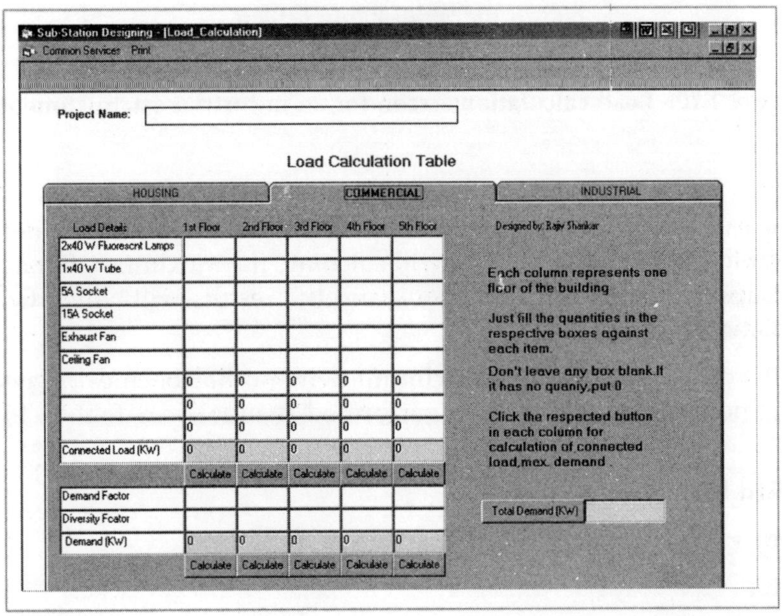

Figure 1.9: Load calculation screen for a commercial establishment

- Refer to Fig. 1.10 Industrial load can be calculated by clicking on the Industrial Tab. The load details, rating of each equipment and total quantity should be filled in the specified column. Click "Calculate" button to calculate the total connected load. The total demand can be calculated by filling in the diversity factor and demand factor and by clicking in the calculate button.

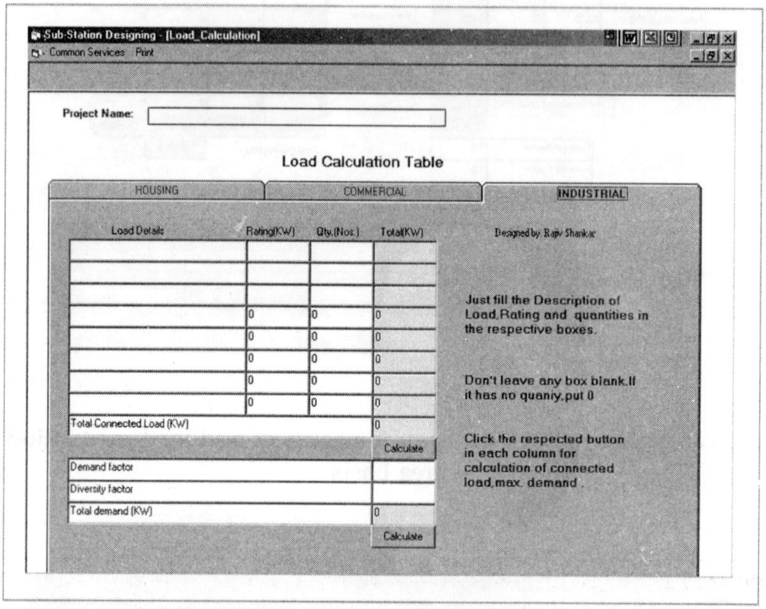

Figure 1.10: Load calculation screen for an industrial establishment

Example 1.8

Let us take the case of calculation of load for a residential complex. The details of load are given in Table 1.4, Example 1.6. We will calculate the maximum demand by using the software supplied along with this book and the results will be verified by doing manual calculation also.

Refer to Fig. 1.7 and Fig. 1.8 where maximum demand has been calculated by using the software. The calculations are also performed manually in Table 1.15 and Table 1.16.

Total connected load = 959.53 KW

Demand factor = 0.7

Diversity factor = 1.4

Total Demand = 947.06 x 0.7/1.4 = 479.76 KW (by Manual Calculation), and 479.7 KW has been calculated by the software (Fig. 1.7).

The Total demand including the common services will be (30.5 KW, table 1.10) + 479.76 KW = 510.26 KW while it comes out to be 510.27 KW by using the software (Refer to Fig. 1.8).

Table: 1.15

Table for total numbers of points:

Sl. No.	Description	Nos. of houses	5 A socket/ house (Nos)	Total (Nos)	15 A socket/ per house (Nos)	Total (Nos)	40 W lamp/ (Nos)	Total (Nos) house	Ceiling fan /house (Nos)	Total (Nos)	100 W lamp/ house (Nos)	Total (Nos)	Exhaust fan/ house (Nos)	Total (Nos)
1.	**BLOCK-1**													
	Type-1	14	12	**168**	05	**70**	07	**98**	05	**70**	04	**56**	01	**14**
	Type-2	14	15	**210**	07	**98**	10	**140**	07	**98**	05	**70**	02	**28**
2.	**BLOCK-2**													
	Type-3	18	18	**324**	09	**162**	11	**198**	09	**162**	05	**90**	03	**54**
	Type-4	18	20	**360**	11	**198**	12	**216**	10	**180**	06	**108**	03	**54**
3.	**BLOCK-3**													
	Type-5	12	22	**264**	13	**156**	15	**180**	12	**144**	08	**96**	03	**36**
	Total	76												

Table: 1.16

Table for calculated load for each type of installation:

Sl. No.	Description	Nos. of houses (Nos)	Total 5 A socket	Total KW (100 W/ socket)	Total 15 A socket (Nos)	Total KW (1000 W/ socket)	Total 40 W lamp (Nos)	Total KW (40 W/ lamp	Total Ceiling fan (Nos)	Total KW (75 W/fan)	Total 100 W lamp (Nos)	Total KW	Total exhaust fan (Nos)	Total KW (100 W/ (KW)	Total Load
1.	**BLOCK-1**														
	Type-1	14	168	**16.8**	70	**70**	98	**3.92**	70	**5.25**	56	**5.6**	14	**1.4**	**102.97**
	Type-2	14	210	**21**	98	**98**	140	**5.6**	98	**7.35**	70	**7.0**	28	**2.8**	**141.75**
2.	**BLOCK-2**														
	Type-3	18	324	**32.4**	162	**162**	198	**7.92**	162	**12.15**	90	**9.0**	54	**5.4**	**228.87**
	Type-4	18	360	**36**	198	**198**	216	**8.64**	180	**13.5**	108	**10.8**	54	**5.4**	**272.34**
3.	**BLOCK-3**														
	Type-5	12	264	**26.4**	156	**156**	180	**7.2**	144	**10.8**	96	**9.6**	36	**3.6**	**213.6**
	Total														**959.53 KW**

2 Transformer

2.0 Introduction

Transformer is an essential part of an electrical system where electrical energy is supplied at high or extra high voltages. Transformers are step-up or step-down type. Power transformers, and generating station transformers are normally step-up transformers while distribution transformers are step-down transformers. The main emphasis in this chapter will be on selection, installation, testing and commissioning of transformer, development of specifications, and other practical aspects. A software has also been provided, which will help in computation of efficiency, load for maximum efficiency, voltage regulation, economics in selection of transformer, and also specifications can be developed by using the software. So, let's get started. First of all we will discuss the various aspects which are considered for the selection of transformer.

2.1 Selection of Transformer

The selection of transformer depends on various factors like construction, installation, various parameters like primary voltage, secondary voltage, power rating, percentage impedance, application etc. Broadly, we can classify the transformers, depending on their features as mentioned in Table 2.1.

Table 2.1

Construction	Selection criteria
Oil-filled transformer with on-load or off-load tap changer. Core and coil assemblies	Oil-filled transformers are the most common type of transformers in power

are kept dipped in transformer oil, which works as a cooling and insulating medium.	and distribution transformers category. These transformers are cheaper than the dry-type transformers. If space is not a constraint and the cost of fire equipment is not much, then these transformers are chosen. It requires a special foundation. Since transformer oil is used, regular maintenance is required. Leakage of transformer oil is the most common fault with these transformers. Further, if variation in voltage is more frequent, then ON load tap changer is provided, because, OFF load type of tap changer will require frequently switching OFF and ON of the transformer, which is not advisable. ON load tap changers are generally used in oil type transformer, since OFF load tap changer for dry type transformer is very costly.
Cast Resin or vacuum impregnated dry- type transformer with OFF load tap changer.	These types of transformers are generally used in residential or commercial complexes, the reason being constraints of space; particularly in big and metropolitan cities space is very costly. Secondly, transformers are installed indoors for aesthetical reasons. These are maintenance-free transformers and very less combustible material is used in the transformers. Due to this, smoke generation is very low. The transformers can be installed near the load.
Installation	
Outdoor or Indoor	Both oil as well as dry types of transformers can be installed outdoors as well as indoors. As mentioned above, in commercial and residential complexes indoor-type of installation is preferred due to safety and aesthetical look, while in industrial establishments outdoor-type installation is more common. If the capacity of the transformer is large then outdoor-type of installation is chosen.
Capacity	Capacity of the transformer is selected, depending on the maximum demand. Besides this, future increase in demand should also be considered. Sometimes,

	mostly in the process industries 100% redundancy is required and more than one transformer is used. Each transformer can take full load. The ratings of the downstream equipments will also depend on the capacity of the transformer. It is important that selection shall be done in a most prudent manner. The excess capacity will result in the increase in rating of all associated equipments and overall increase in cost of substation.
Primary and secondary voltage	Transformer is chosen as per the receiving end and delivery end voltage. Accordingly, it may be step-up or step-down transformer.
Percentage impedance	Percentage impedance is one of the most important aspect of a transformer. Voltage regulation and short circuit current depend on it. If percentage impedance is low, the short circuit current will increase and the short circuit rating of the associated equipments will also increase resulting in overall increase in cost, while the regulation of low percentage impedance transformer will be good. So a trade-off is made between the two, i.e. short circuit current and voltage regulation.
Economics	
Losses	Cost of core losses, load losses and auxiliary losses shall be taken into account along with the capital cost while evaluating the transformers.
Efficiency	For fixed type of load, transformer should be chosen in such a way that it operates at maximum efficiency.
Type of load	Type of load, like furnace load, motor load, lighting load etc., should also be considered while selecting a transformer. Lighting load is normally unbalance load and it should be segregated from other types of load. Unit ratio transformer is provided for lighting load.

2.2 Comparison between Dry-Type and Oil-Filled Transformer

The various parameters of dry-type and oil-filled transformers are compared in Table 2.2.

Table 2.2

	Oil-filled transformer	Resin-impregnated dry-type transformer	Cast-resin dry-type transformer
Construction	The core-coil assembly is put in the tank filled with transformer oil. Transformer oil works as a cooling as well as insulating medium. Conservator is provided to accommodate the expansion of transformer oil due to heating during operation of transformer. ON-load or OFF-load tap changer can be provided.	The LV and HV windings are cast impregnated under high vacuum. Impregnated and cured windings are assembled on core limbs. For indoor installations enclosure of IP-33 degree of protection and for outdoor installations enclosure of IP-45 degree of protection are normally provided. The dry-type transformer is generally fitted with OFF-load tap changer. Insulation Class of the transformer is F, and it can withstand (90^{0}C) above ambient temperature.	The main difference between resin-impregnated and cast-resin type of transformer is that the LV and HV windings are cast in resin under vacuum.
Installation	Transformer can be installed indoors as well as outdoors. A special concrete foundation is required. If the quantity of oil is more than 2000 litres, then as per India Electricity Rules, soak pit is required. Further fire wall of two-hour rating between the two transformers is required.	Transformer can be installed indoors as well as outdoors. No special foundation is needed, since its weight is comparatively less. It can be installed near the load so that the cost of cabling is reduced.	Transformer can be installed indoors as also outdoors. It can be installed near the load, and no special foundation is required.

Maintenance	It requires regular maintenance like filtration of transformer oil, testing of various characteristics of transformer oil like viscosity, dissolved gases, di-electric strength etc.	No maintenance is required.	No maintenance is required.
Operational safety	If not maintained properly, it may cause fire hazard. Smoke generation is heavy and toxic. It has sufficient quantity of combustible material.	No fire hazard. It has very less combustible material. Smoke and toxic gases generation is very low.	Smoke generation is low. The resin is non-combustible and self-extinguishable.
Cost	Two and half times cheaper than dry type transformer.	More	More
Operating temperature	80^0C, maximum	180^0C, maximum	130^0C, maximum
Cooling	The core-coil assembly remain immersed in transformer oil; and transformer oil works as a cooling and insulating medium.	The dissipation of heat is through convection and radiation only. Higher hot spot temperature.	The dissipation of heat is through conduction, convection and radiation.
Environment	No additional requirement, breather takes care of the humidity in the air.	Requires additional dehumidifiers, space heaters.	Non-hygroscopic and can work in humid/ highly humid atmosphere.
Noise Level	High	High	Less
Repair	Winding can be repaired.	Winding can be repaired.	Winding cannot be repaired.

2.3 Specifications

For procurement of a transformer, technical specifications are prepared, so that the manufacturer can quote the rates based on those specifications. The parameters to be specified in the enquiry issued to the vendors are mentioned below:

- Type of transformer
- Quantity required
- Rated KVA
- Frequency
- No. of phases
- Voltage ratio
- Winding connection
- Winding material
- Tapings
- Type of tap selector
- Requirement of neutral terminal if winding is connected in star
- Phase connection
- Cooling
- Location of transformer
- Permissible overload limit
- High and low voltage terminal arrangement
- Ambient temperature
- Impulse level
- Temperature rise

  Some of the parameters are specified by the vendor along with their offer like.

  a. Efficiency at specified power factor and load

  b. Load losses at rated load and specified temperature

  c. Core losses at rated voltage

- Accessories required:

  a. Oil and winding temperature indicator

  b. Oil and winding temperature alarm and tripping switch

  c. Oil conservator

  d. Buchloz relay

  e. Rollers

In addition to the above, the other information like tests required to be performed, any special test, witnessing of tests at manufacturer's works and documents to be submitted are also specified in the technical specification.

2.4 Installation

For installation of transformer, these should be followed:

- Manufacturer's recommendations
- IS:10028-1981, Part II, Code of Practice for Selection, Installation and Maintenance of Transformers.
- The Indian Electricity Rules-2003

If the transformer is going to be installed indoors, then the question arises: what should be the size of the room for accommodating a transformer of a particular rating? The most important thing is that proper ventilation should be ensured, i.e. free movement of air around the four sides of the transformer.

The minimum clearance between the highest point of the conservator tank to the ceiling of the transformer room should not be less than 0.5 metre. The minimum spacing between the wall and transformer should be as per IS:10028 (Part-2) which is mentioned below:

- Wall on one side: 0.5 metre
- Wall on two sides: 0.75 metre
- Wall on three sides: 1.00 metre
- Wall on four sides: 1.25 metre

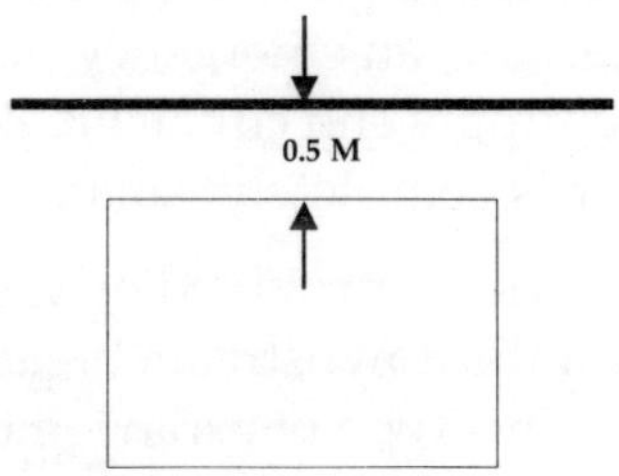

Distance from wall 0.5 metre

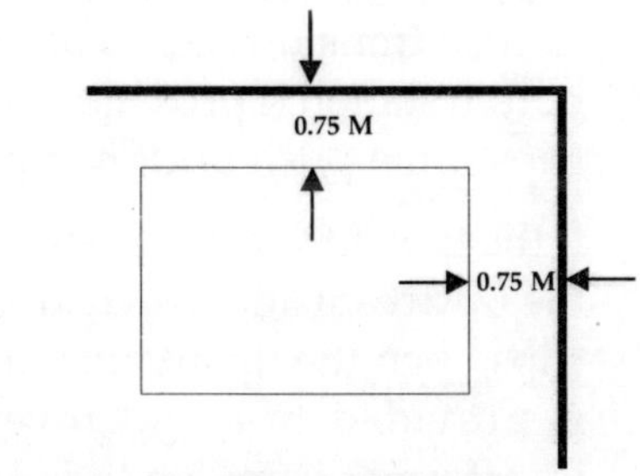

Distance from wall 0.75 metre

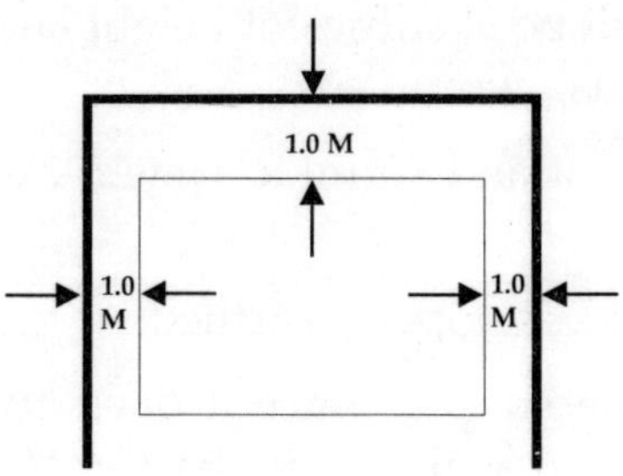

Distance from wall 1.0 metre

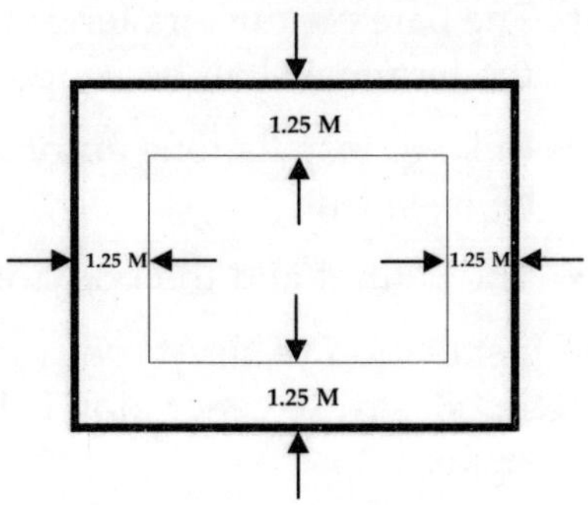

Distance from wall 1.25 metre

Figure 2.1: Minimum spacing between the wall and transformer

For lifting the transformer core at a later date for maintenance purposes it would be a good idea to provide a lifting beam secured with the structure. Sufficient space should be provided for taking oil sampling and also for connecting the machine for dehydration etc. in case of oil-type transformer.

A level concrete plinth, above the flood water level shall be provided for installation of transformer. In case of outdoor installation, the cable boxes (if provided), bus-ducts (if provided) shall be sealed and a canopy shall be provided to prevent entry of moisture.

2.5 Safety Requirements

- Transformers installed inside the buildings where the oil capacity exceeds 2000 litres of oil shall be provided with oil soak pit. Oil soak pit shall be of RCC. Oil soak pit may be provided either under the transformer or outside the transformer room.
- If two or more transformers are installed side by side, they shall be separated by fire-separation walls of two hours rating. Fire separation wall is required even if the oil capacity of the individual transformer does not exceed 2000 liters and total capacity of transformer oil of all the transformer installed side by side exceeds 2000 litres.
- Means shall be provided for complete isolation of every transformer from the supply and these shall be so placed as to be readily accessible to enable the supply to such transformers to be cut off immediately. For example, an emergency push button station is provided near the transformer and the supply end circuit breaker trips if the push button is pressed and the transformer is completely isolated.
- If the transformers are operated in parallel, then an interlock is provided between the upstream and downstream circuit breakers so that if the downstream breaker trips, then the upstream side breaker should also trip. This type of arrangement is provided to avoid reverse feedback of power supply and to isolate the transformer completely at both ends.
- The bare conductors terminated at the bushings shall be at sufficient height and the terminal shall be properly protected against birds, vermin etc.
- In large installations automatic firefighting equipment and sprinkle system can be installed.
- The neutral and the body of the transformer should be properly earthed.

The installation of dry-type transformers is comparative simple. Since it does not contain oil, smoke generation is low. Therefore no special oil soak pit, fire barrier etc. are required.

2.6 Commissioning

After installation of transformer and oil filling, dehydration is carried out as per the specified procedure. Before charging of transformer, the following tests are carried out.

- Ratio and Polarity Test

  The ratio test is done with the help of ratio meter. However, at site normally ratio meter is not available and it can be done by applying low voltage at HT end, but this method is not very accurate.

  For checking polarity or phaser group, 415 volts is applied at HT side and the voltages between all the phases on low voltage side is measured at every tap position. To check phasing, the R and r terminals are short circuited and voltages between Y-y, Y-b, B-b and B-y are measured. For Dyn 11 Connection voltage between Y-y, Y-b should be same and less than B-y.

- Insulation resistance.

  Megger is used to check the insulation resistances between the phases and earth and between phases and also between HV and LV phases.

- Winding Resistance.

  DC resistance is measured by Voltmeter/Ammeter method or by micro ohm meter.

- No load current and Core Balance Test.

  The low voltage is applied on LV side the HT side is open circuited. The magnetizing current is measured.

- Tripping and alarm circuit tests, Buchloz relay test, tap changing mechanism test, tests for proper operations of cooling fans and oil circulation pump are some of the tests to be conducted before commissioning of the transformer.

2.7 Maintenance

The important part of a transformer is the transformer oil. The transformer oil should be checked at regular intervals. The testing of the transformer oil reveals the healthy state of other parts of the transformer too, like core and winding. Since core and winding are immersed in the transformer oil, any abnormality in core and winding also affects the transformer oil. For example, any internal short-circuiting causes disintegration of transformer oil and generation of gases. Manufacturer should also be consulted about the maintenance to be carried on the transformer. The other most common type of defect is the leakage of transformer oil from gasket or any other part, which should be attended to at the earliest.

The following tests are performed on the transformer oil:

- Dielectric strength
- Water content
- Neutralization value (Total acidity)
- Interfacial tension
- Dielectric dissipation factor (Tan delta)
- Resistivity
- Flash point
- Sediments
- Dissolved gases analysis

In Table 2.3 the significance of each test and the standard values are given.

Table 2.3

The dehydration and filtration of transformer oil shall be carried out once in a year. During dehydration, the insulation resistance shall be measured at regular interval.

Test	Significance of test	Indian Standard	Standard Value
Dielectric strength	This test indicates the resistance of oil to the electric stress. The electric stress is affected by the moisture and also polar material.	IS: 6792 IS: 1866	>30KV for one minute
Water content	It affects the dielectric strength and can cause arcing and subsequently failure of transformer.	IS: 13567 IS: 1866	≤20ppm
Acidity	The acidity test is used to determine the chemical aging of the oil. The acid formation is due to the oxidation of the oil, and it actively encourages deterioration of the insulating papers and press boards used in the manufacturing of transformer.	IS:1448 IS: 1866	0.300 mg KOH/gm Maximum
Interfacial tension	Interfacial tension of oil against water is determination of presence of polar contamination in oil. Contamination by polar material accelerates oxidation of oil and subsequently overheating and failure of transformer.	IS: 6104	15 mn/m Minimum

Dielectric dissipation factor (Tan Delta) at 90°C and at 40 to 60 Hz	It is used to determine the soluble contamination and the power loss in the transformer	IS: 6262	1.000 Maximum
Resistivity At 90°C	Decreased resistivity indicates presence of contamination, water etc. in the oil	IS: 6103	0.100×10^{12} ohm cm Minimum
Flash point	It is the lowest temperature at which the vapour of the oil gets ignited. Insulating oil undergoes chemical changes during operation and the flash point may get lowered. Low flash point results in fire.	IS: 1448	140°C Minimum
Sediments	Sediments reduce the electric strength of the oil and the heat exchange is hindered, which further deteriorates the insulating material.	IS: 1866	<0.1 (percentage by mass) Maximum
Dissolved gas analysis	The fault in transformer like arcing, overheating etc. results in chemical changes in the oil due to which the gases are released. The incipient fault of low intensity results in releases of gases, which gets dissolved in the oil. The analysis of the dissolved gas can help in taking preventive action much before the fault takes place. The most significant gases generated by decomposition of oil are hydrogen (H_2) methane (CH_4), ethane (C_2H_6), ethylene (C_2H_4) and acetylene (C_2H_2). Carbon dioxide (CO_2) and carbon monoxide (CO) are produced at normal operating temp and as a result of normal ageing of insulation i.e. gradation of cellulosic materials takes place. If CO_2/CO is below 3 or above 11, it should be regarded as a fault involving cellulose.	IS: 9434 and IS: 10593	Hydrogen: 5 ppm, (sensitivity vlaue) Carbon oxides: 25 ppm (sensitivity vlaue) Hydrocarbons: 1 ppm (sensitivity vlaue) Atmospheric gases: 50 ppm (sensitivity vlaue) Gas present in significant quantity is the level at which it is at least two times the sensitivity or detection limit. For investigation of fault in the transformer the gas contents should be compared with the initial values of gas contents, when the transfromer was commissioned.

The colour of silica gel in the breather should be checked, and the silica gel should be replaced whenever required. All the safe devices like Buchloz relay, winding and oil temperature tripping and alarm circuit shall be checked at regular interval.

2.8 Other Important Aspects of Transformer

Significance of percentage impedance

When selecting a transformer, percentage impedance is one of the important parameters, which should be taken care of. To keep the voltage drop small, low percentage impedance is required; but low percentage impedance increases the fault current. Therefore, according to the application of transformer, a compromise is made between the voltage drop and the fault current.

The percentage impedance of a transformer is the percentage of the rated primary voltage applied to the transformer to produce the normal rated full load current in its short-circuited secondary. Percentage impedance is the percentage voltage measured as impedance.

Percentage impedance (Z) = $\sqrt{3}(Ez \times 100)/En$

En = Normal rating of primary voltage

Ez = Applied primary voltage to produce normal rated current when secondary is short-circuited.

Inrush current

When a transformer is energized, a large amount of current, approximately 7 to 8 times the rated primary current, flows through the transformer primary winding for a fraction of a second. Sometimes it causes tripping of the over-current relay due to creation of fault-like conditions. The reason of this current inrush is explained below.

The magnitude of the inrush current depends on two factors:

1. Voltage at the instant of closing of the transformer circuit breaker.
2. Residual magnetism left in the core.

We will examine the six conditions under which inrush of current takes place.

1. The voltage is zero at the instant of switching in and the residual magnetism is zero too.
2. The voltage is zero and the maximum residual magnetism is present; but the polarity is opposite to the flux, which will build up under normal conditions.

3. The voltage is zero and the maximum residual magnetism is present; but the polarity is same to the flux, which will build up under normal conditions.
4. The voltage is maximum at the instant of switching in and the residual magnetism is zero.
5. The voltage is maximum and the maximum residual magnetism is present; but the polarity is opposite to the flux, which will build up under normal conditions.
6. The voltage is maximum and the maximum residual magnetism is present; but the polarity is same to the flux, which will build up under normal conditions.

Let us consider the first condition. The magnetic flux lags ninety degree out of phase with the voltage. This means that the flux reaches its maximum value when the voltage is zero and it becomes zero when the voltage is at its maximum value. Therefore, the flux vary to its maximum value in one direction and to maximum value in the opposite direction to produce the half cycle of back emf in the primary winding. Therefore, the total flux density in the core will be approximately double the normal value.

The value of no load current also varies with the flux density, and it may be many times more than the normal current.

Under the second condition, the flux will not start from zero. But since the residual flux of opposite polarity is also present, it will start from a value corresponding to the magnitude and polarity of the residual magnetism and the flux density during first cycle will be higher than the previous condition, and so the magnitude of the inrush current will also be higher.

The third condition will be opposite to the second one. No inrush of current will take place, if the residual magnetism is equal to the maximum flux density. If the value of the residual magnetism is lower than the maximum flux density, then due to unsymmetrical conditions of the flux wave, inrush of current will take place.

Under the fourth condition, the flux will be zero when the voltage is maximum due to ninety degree phase displacement. Residual magnetism is also zero. There will be no current inrush.

Under the fifth condition, due to the presence of residual magnetism of the opposite polarity, flux wave will be unsymmetrical to the zero axis and current inrush will take place.

Under the sixth condition, again current inrush will take place due to presence of residual magnetism of the same polarity. The flux wave will not be symmetrical as mentioned in the fifth condition.

Abnormal operating conditions

Table 2.4

Sl. No.	Condition	Effect
1.	Lower frequency than the designed value	Induced emf will be low. Output voltage will be low. The flux density will increase. Hysteresis losses will increase. Efficiency will get reduced.
2.	Higher frequency than the designed value	Induced emf will be high. Output voltage will be high. Eddy current losses will be high. The efficiency will get reduced.
3.	High input voltage	Flux density will increase. Core losses will increase. Output voltage will get increased proportionately. Insulation will be stressed. Life of transformer will be reduced.
4.	Low input voltage	Flux density will get decreased. Output voltage will get decreased. If secondary is connected with motor load, the secondary winding current will get increased.
5.	Harmonics	Third harmonic current is normally present in the transformer windings due to non-linear characteristic of the hystersis loop. In a distribution transformer with Delta connected primary winding and star connected secondary winding with neutral, the third harmonic current circulates in the delta windings and it will not be reflected in the line, while at LT side it circulates through neutral, line and transformer phase winding.

2.9 Evaluation of Transformer Losses

While placing order on the transformer, it is necessary that the losses of the transformers should also be evaluated and added to the capital cost of the transformer. The cost of losses for each transformer should be evaluated by multiplying the load losses at 75^0C by appropriate Rs/KW and no load losses at 100% voltage. In addition to this, a penalty should also be imposed on the manufacturer/supplier of the transformer if the actual tested transformer losses exceed the guaranteed losses. The transformer has three types of losses.

- Load losses
- No-load losses
- Auxiliary losses

The no-load losses are hystersis, eddy current losses and the I^2R losses in the winding due to the excitation current. The load losses are the I^2R losses in the winding. The auxiliary losses are the power required to run the cooling fans, pumps etc.

We will discuss this in more detail in the following worked example.

Example 2.1

Selection of transformer

In Chapter 1, the total demand for LT system of steam generation plant works out to be 887 KW. For choosing the transformer, the following steps are taken.

Step 1

Before the selection of the transformer, the following points are to be taken into consideration:

- The design philosophy i.e. how critical is the load supplied by the transformer.
- Incoming supply voltage and variations in the incoming supply voltage.
- How much overloading of the transformer is allowed.
- Location where the transformer will be installed.

As mentioned in Chapter 1, usually, for a continuous process plant, 100% spare capacity is required, so that in case of failure of one transformer, the other transformer can take the full load. Therefore, to cater for a load of 887 KW, two transformers are required. General practice is to select the equipment of equal rating, so that the inventory of spare parts gets reduced. Therefore, two transformers of equal rating will be chosen and load will be divided in such a way that each transformer will take half of the load i.e. 466.84 KVA (Total load is 933.68 KVA, assuming power factor as 0.95). Two numbers of 1000 KVA transformers are chosen. Normally, the transformers operate on maximum efficiency at 40–50% of its rated capacity. Further, in case of failure of one transformer, the other transformer of 1000 KVA will take the entire load.

Step 2

The specifications are developed for procurement of transformer. The 887 KW load will normally be supplied at 11 KV. If the variation in the incoming supply is more frequent, then it will be better option to have ON-Load tap changer type of transformer, because in case of OFF-Load type, it is not advisable to switch off the transformer frequently and change the Tap. If the transformer is to be located indoor, then dry-type transformer is preferable. In that case, the cost of the cable will also get reduced, since the switch gear can be installed very near to the transformer. There are various other advantages of the dry-type transformers, which are already mentioned in this chapter.

Step 3

The tender is prepared as per the specifications and sent to the transformer manufacturers along with the commercial terms and conditions like terms of payment, delivery period, guarantee etc.

Step 4

Now the bids are evaluated and the most important factor in bid evaluation is to take into account the cost of no-load losses and load losses along with the cost of the transformer. Now the question arises: how to evaluate the cost of transformer losses. While preparing the bids, the per KW cost of load losses and no-load losses should be specified so that the vendor should have the prior information that how much cost towards losses will be added in their price bid at the time of evaluation of bid.

Evaluation of cost of transformer losses

We assume that the two bids are received from the transformer manufacturers.

Bid A

Capital cost: Rs 3,50,000

Load losses: 12500 W

No load losses: 1800 W

Bid B

Capital cost: Rs 3,80,000

Load losses: 11800 W

No load losses: 1600 W

Life of both of the transformers is 20 years.

Other data available with us are as mentioned below:

- Load factor = 0.47
- Demand factor = 0.67
- Interest rate = 7% of capital cost
- Insurance rate = 0.6% of capital cost
- Depreciation = 3% of capital cost
- Operation and maintenance charges = 1% of capital cost
- Demand charges = Rs 250/KW per month (Rs 3000/KW per year)
- Energy charges = Rs 4.5

Fixed charge rate = 7% + 0.6% + 3% + 1% = 11.6%

Transformer loading per year = 8760 Hrs.

Cost of no-load losses per KW

Demand charges (annual) + Hours (Yearly loading) x Unit charges

= 3000 + (8760 x 4.5)

= Rs. 42,420.00

Cost of copper losses per KW

Loss load factor = 0.8 x Load factor^2 + 0.2 x Load factor

= 0.8 x 0.47^2 + 0.2 x 0.47

= 0.27

Cost of losses per KW = Demand factor ^{2} x (demand charges + (8760 x unit charges x Load loss Factor)

= 0.67^2 x (3000 + (8760 x 4.5 x 0.27)

= Rs. 6,124.52

Evaluation

	A (Rs.)	B (Rs.)
Cost of no-load losses	42420 x 1.8 = 76,356.00	42420 x 1.6 = 67,872.00
Cost of load losses	6124.52 x 12.5 = 76,556.52	6124.52 x 11.8 = 72,269.36
Bid price	3,50,000.00	3,80,000.00
Total	5,02,912.50	5,20,141.36

Annual running cost

Transformer A

(Cost of Transformer x carrying charges/100) + (Cost of load losses + Cost of no-load losses)

(350000 x 11.6/100) + (76356 + 76556.5)

= Rs.19,3,512.50

Transformer B

(380000 x 11.6/100) + (67872 + 72269.36)

= Rs.18,4221.36

Present value of running cost

N = Life of transformer in years.

P = (1+interest on capital/100) ^{N}

Present value of running cost = Annual running cost x (100/Interest on capital) x (1-1/P)

Transformer A

P = (1+ 7/100)^{20}

= 3.87

Present value of running cost = 19,35,12.5 x ((100/7) x (1-1/3.87))

= 20,50,132.4

Transformer B

Present value of running cost = 184221 x ((100/7) x (1-1/3.87))

= 19,51,695.5

Total cost of Transformer A

20,50,132.4 + 3,50,000

= Rs. 24,00132.4

Total cost of Transformer B

19,51,659.5 + 380000

= Rs. 23,31,659.50

Step 5

Now, the final step is to develop technical specifications for procurement of transformers.

- Rated KVA: 1000
- Type of transformer: Cast resin dry type
- Location: Indoors

- Maximum ambient temperature: 45⁰C
- Number of phases: 3
- Frequency: 50 Hz
- Winding connection: Delta-Star
- Winding material: Copper
- HV and LV terminal arrangement: HV cable and LV cable
- Percentage impedance: 4.5%
- Permissible overload limit: 10%
- Type of cooling: AN (Air Natural)
- Phasor diagram: Dyn11
- Neutral terminal: Required
- Voltage ratio: 11 KV/433 V
- Tapings: ±5%
- Tap selector: OFF-Load type
- Winding temperature indicator: required
- Marshaling box: Required
- Tests: All routine tests
- Inspection: Shall be done at manufacturer's place
- Documents: All test certificates and operating manuals

Information to be submitted by the vendor:

- No-load losses at rated voltage
- Load losses at rated current
- Efficiency at unity power factor and at 125%, 100%, 75%, 50%, 25% load
- Regulation at unity power factor and 0.8 power factor and rated load
- Temperature rise at rated output
- Weight of transformer

2.10 Conclusion

After going through the chapter we will be able to carry out the following tasks:

- Choosing a transformer of correct type and rating for a specified application
- Evaluating the cost of the transformer

- Preparation of technical specifications
- Installation, testing and commissioning of transformer

In this chapter, emphasis is given on the practical aspects of the transformer like comparison between dry and oil type of transformers, installation, testing and commissioning, oil testing. Significance of other important parameters like percentage impedance, inrush current during switching on of the transformer etc. are explained. Detailed theoretical aspects are not covered for which many books are already available.

2.11 Software

Objective

- Selection of a transformer.
- To calculate the load factor at which the efficiency is maximum.
- To calculate the efficiency for a specified load and power factor.
- To calculate the voltage regulation/voltage drop of a transformer for the specified load factor and power factor.
- To calculate the annual capitalized cost of the transformer and on that basis selects the most economical transformer.
- To prepare the specifications of the transformer and to print the same.
- To store, retrieve and delete the parameters of transformer in a library.

How to use the software

- Click on the "Electrical Equipment" from the main menu.
- Click on "Transformer" from the sub menu; refer to Fig. 2.2. The transformer screen opens up Fig. 2.3.
- Fill in the maximum demand in KVA, numbers of transformers to be provided to cater for the above load like 1, 2 or 3 as the case maybe and KVA rating of the transformer, in the respective boxes.
- Click the "Calculate Load Factor" button to calculate the load factor.
- Enter the "No Load" and "Load Losses" in KW.

- Click the button "Load Factor for Maximum Efficiency". The purpose of this calculation is that the load factor of the chosen transformer should be near to the calculated load factor so that the transformer should operate at maximum efficiency considering that the load on transformer remains constant.
- Enter the power factor and click the "Efficiency" button to calculate the efficiency of the transformer at that power factor and calculated load factor.
- The above steps can be repeated for finding the efficiency of different ratings of transformers.
- To calculate the regulation, enter the percentage impedance, power factor and the load factor in the respective boxes at which the regulation or voltage drop is required to be calculated.
- Click the "percentage voltage drop" button to calculate the voltage drop, or regulation of the transformer.

It is assumed that the load is equally distributed between the transformers. Refer to Fig 2.3 and Example 2.2, 2.3, 2.4 and 2.5.

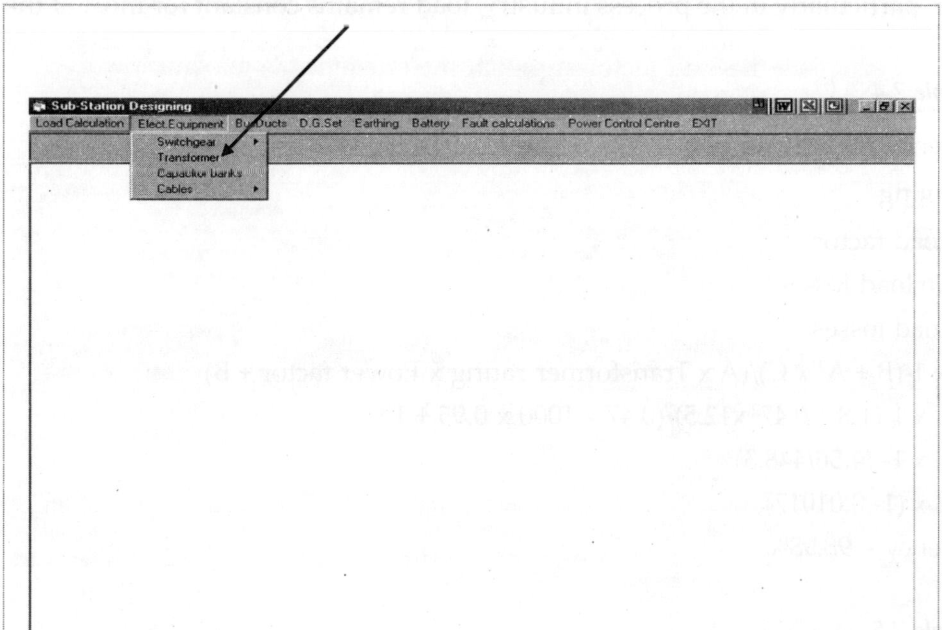

Figure 2.2: Screen showing main menu bar and sub menu

Let's take an example.

Example 2.2

Maximum demand: let's take 945 KVA

Number of transformers to be provided: 02

Rating of each transformer: 1000 KVA

Load factor: Maximum demand/(KVA rating of Transformer x numbers of transformer)

Therefore load factor = 945/(1000 x 2) = 0.47

Example 2.3

Suppose no-load losses are 1.8 KW and load losses are 12.5 KW, then the load factor for maximum efficiency would be: √(No Load losses/Load losses) = √(1.8/12.5) = 0.38.

The idea of calculation of load factor for maximum efficiency and also for maximum demand is that both load factors should be near to each other so that the transformer operates on maximum efficiency, provided the load remains constant. In industrial sector, particularly in the process industry, load remains constant for most of the time.

Example 2.4

Efficiency for a given power factor and load factor will be:

Assuming

A = Load factor

B = No load losses

C = Load losses

$100 \times 1-(B + A^2 \times C)/(A \times \text{Transformer rating} \times \text{Power factor} + B)$

$= 100 \times 1-(1.8 + 0.47^2 \times 12.5)/(0.47 \times 1000 \times 0.95 + 1.8)$

$= 100 \times 1- (4.56/448.3)$

$= 100 \times (1- 0.01017)$

Efficiency = 98.98%

Example 2.5

Percentage resistance voltage = Load losses x 100/Transformer rating

= 12.5 x100/1000 = 1.25% = Va

Percentage reactance voltage = $\sqrt{(\%\text{impedance}^2 - Va^2)}$ = Vb

$= \sqrt{(4.5^2 - 1.25^2)} = 4.32\%$ = Vb

Percentage Voltage Drop = Load factor x (Va x cosϕ + Vb x sinϕ) + Load factor ^{2} x (Vb

$x \cos\phi - Va \times \sin\phi)^2 /200$

$= 0.47 \times (1.25 \times 0.95 + 4.32 \times 0.31) + 0.47^2 \times (4.32 \times 0.95 - 1.25 \times 0.31)^2/200$

$0.47 \times 2.526 + 0.0152 = 1.202\%$

Refer to Fig 2.3, which shows the computer screen where the above calculations were done by the software.

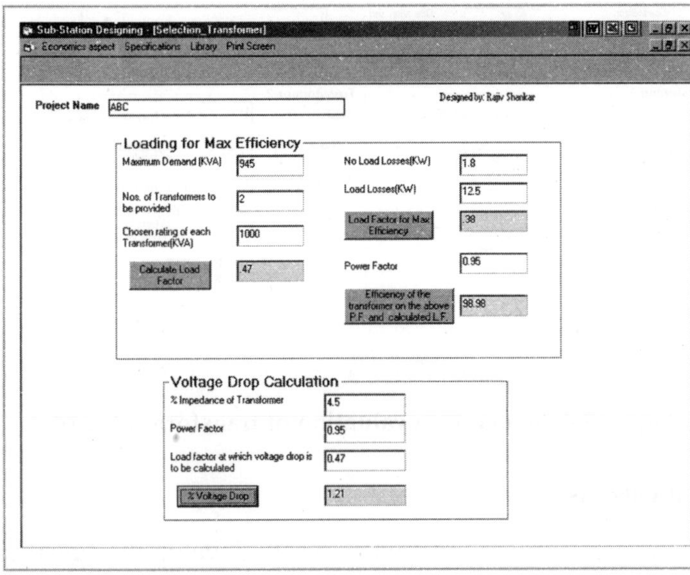

Figure 2.3: Screen for calculation of load factor for maximum efficiency and voltage drop

Economics aspects

To look into the economics aspects in selection of a transformer, the annual running cost including the capitalized value of the losses is taken into account.

- Click on the "Economic Aspect" on the menu bar of the transformer screen.
- A new screen opens up, which compares the annual running cost of the two transformers.
- Enter the data which is common for both the transformers in the respective boxes provided in common data frame.
- Calculate the load loss factor and fixed carrying charges by clicking the respective button.
- Enter the load and no load losses in KW for both the transformers in the respective boxes.
- Cost of load losses, no-load losses and total cost transformers are calculated by clicking the respective buttons. Refer to Example 2.1 and Fig. 2.4.

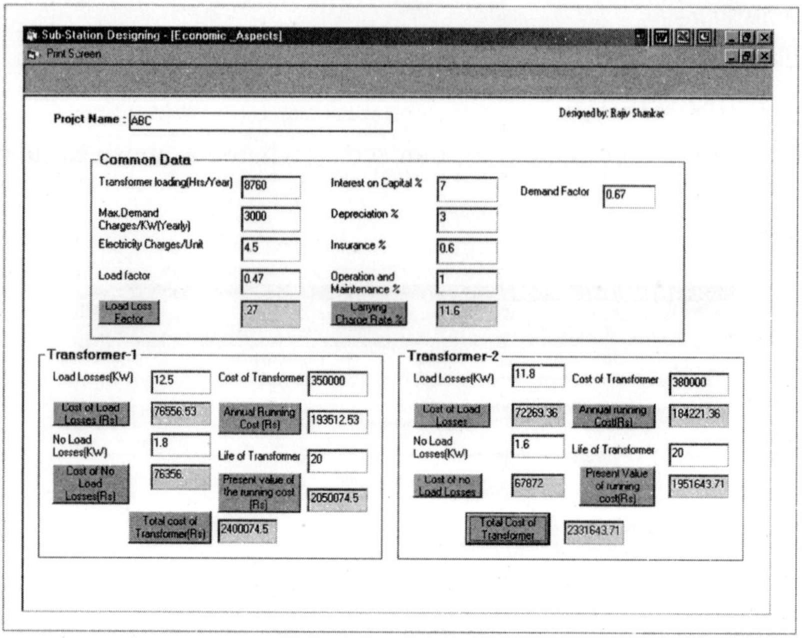

Figure 2.4: Screen for evaluation of transformer's cost

Developing specifications

After selection of the transformer, the next step is to develop the specifications. Follow the following steps for developing the specifications.

- Click on "Specification" on the main menu bar of the transformer screen.
- A screen opens up. A number of boxes and list boxes are provided. Information stored in the list boxes can be chosen or if the information is not available in the list box, it can be manually filled in the boxes. However, "ADD" button should be clicked before filling in the boxes.
- By clicking on "Print" on the menu bar the specification screen can be printed. Refer to Fig. 2.5
- The specifications can also be generated in a tabulation form by clicking on the "Generate Specification" on the menu bar.

This specification can also be printed by clicking on the "Print Button" on the menu bar (Refer to Fig. 2.6).

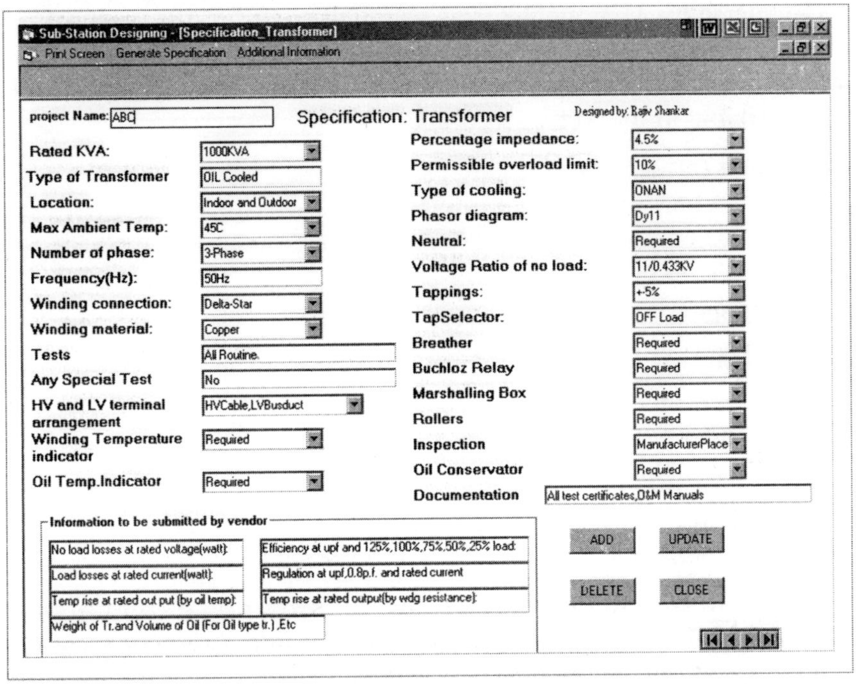

Figure 2.5: Screen for generation of specifications

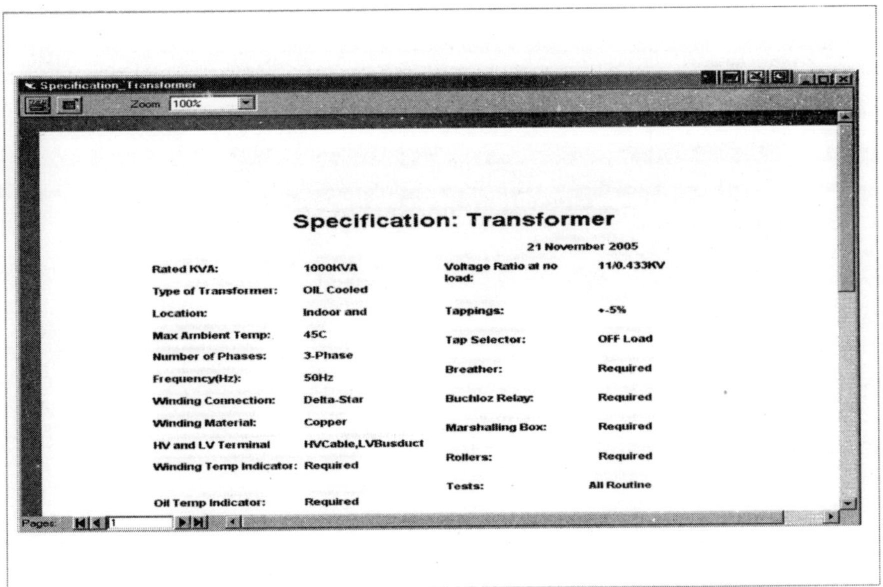

Figure 2.6: Generation of specifications

Library

The software has also got the facility to develop a library of various parameters of the transformer available in market.

The following data can be stored.

- Make
- Rating
- Primary voltage
- Secondary voltage
- Type
- Load losses
- No load losses
- Percentage impedance

To store the data in the library click on the "Library Entry" on the menu bar of transformer screen. Click "Add" button and enter the data. After entry click "Update". The entered data is stored in the library. Refer to Fig. 2.8

To refer library click on the "Library" button on the menu bar. A table containing all the above mentioned data opens up. Refer to Fig. 2.7.

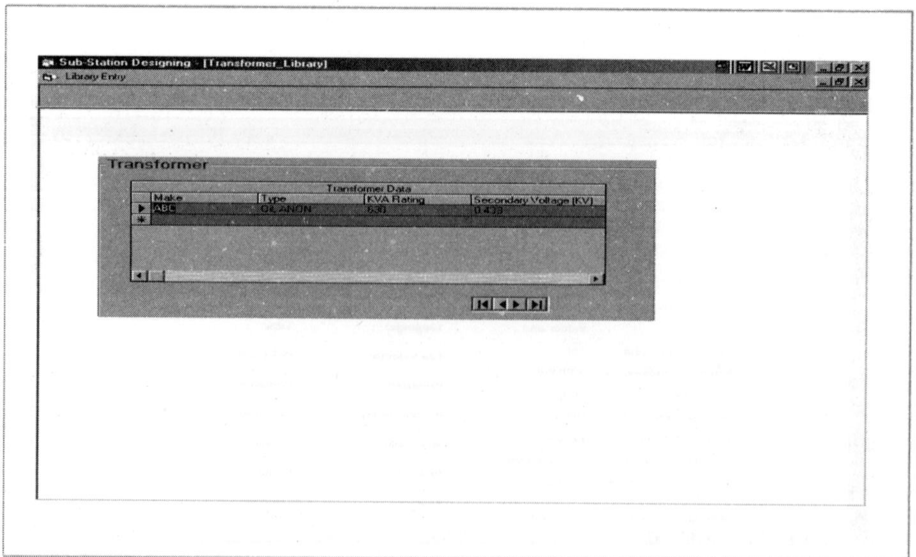

Figure 2.7: Screen for retrieving information from the library

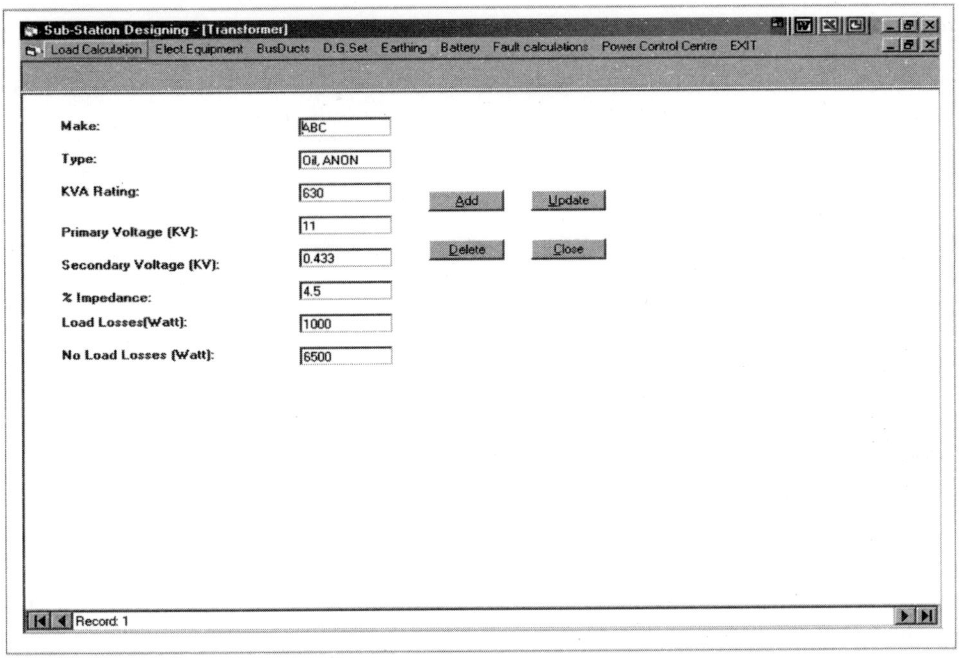

Figure 2.8: Screen for making and deleting entry in the library

3 Power Control Centre

3.0 Introduction

Power Control Centre (PCC), Motor Control Centre (MCC), Low-Tension (LT) switch-gear, High Tension (HT) switch-gear, these are the words frequently heard whenever reference of any substation is made. Basically, all have metal-clad housing and consist of bus-bars and circuit breakers, measuring and protection equipments. The motor control centre particularly has motor starters consisting of moulded case circuit breakers, switch fuse units, contactors, overload relays, single phase preventer etc. The starters may be direct-on-line type or star-delta or electronic (soft starters). The type of starter depends on the design philosophy, application etc.

In this chapter we would highlight specifications, installation, testing of LT/HT switchgear, and finally future trends in the switchgear.

We will also look into the components of the metal clad switch-gear like bus-bars, circuit breakers etc.

It is necessary to prepare a single-line diagram of an electrical system before preparing specifications of a Power Control Centre. The single-line diagram shall indicate the bus-bar scheme, rating of bus-bars, arrangement of incoming and outgoing feeders, protection scheme including protection relays, instrument transformers etc., electrical parameters measurement scheme, interlocking of upstream, downstream feeders and incomers and bus couplers, indications, alarms and annunciation, load on each feeder, rating and sizes of cables, rating of bus-duct (if provided).

3.1 Bus-Bar

Scheme

It is important to do a little bit of planning in the beginning before finalization of the scheme of the distribution system. While deciding a particular arrangement, care should be taken that the arrangement should be flexible, easily maintainable and economical.

Single bus-bar

This is the simplest and economical scheme (Refer to Fig. 3.1). It has only one bus-bar. Incoming as well as outgoing feeders are connected to this bus. If fault takes place on the bus, total system has to be shut down and power supply cannot be restored quickly. This type of system is not flexible. Where economics is the main consideration and load is of non-critical nature, this type of system can be used. Further, for small substations where HT/LT power is supplied with a single feeder, this type of arrangement is used.

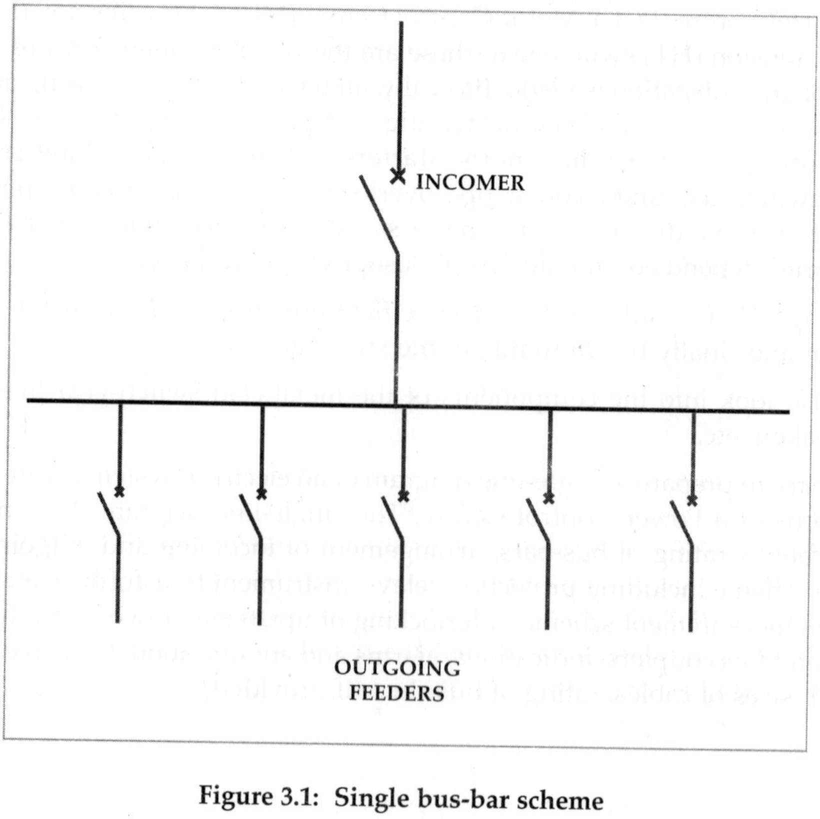

Figure 3.1: Single bus-bar scheme

Sectionalized bus-bar

In some industries like the process industry, the availability of power supply is very critical and disruption of power supply for a longer duration causes a very adverse effect on the productivity of the industry. The restoration of supply in shortest possible time is required. In such type of situation 100% redundancy of power supply is desirable. The most common system is sectionalized bus-bar arrangement with more than one incomer (Refer to Fig. 3.2). A simple system has two incomers with one bus-coupler. In normal condition, the bus-coupler remains open while the two bus bars are fed independently by two incoming feeders. Each feeder is capable of feeding the entire load of the PCC in case of failure of power supply of either of the incoming feeder. The bus coupler is used to connect the two buses when the system is fed by one source. Sometimes, to restore the power supply quickly, auto changeover scheme is implemented i.e. the bus-coupler closes automatically on failure of power supply of either of the incoming feeder. To prevent synchronization of the two incomers, the bus-coupler is electrically interlocked with the two incomers, so that at a time all the three circuit breakers could not be closed. Further, changeover of the supply will only take place if the bus section of which the incomer has tripped is healthy. This is a simple scheme of two incomer and one bus-coupler. The PCC can have more than two incomer and more than one bus-coupler with auto changeover scheme. The sectionalized system is more flexible in comparison to the single-bus system.

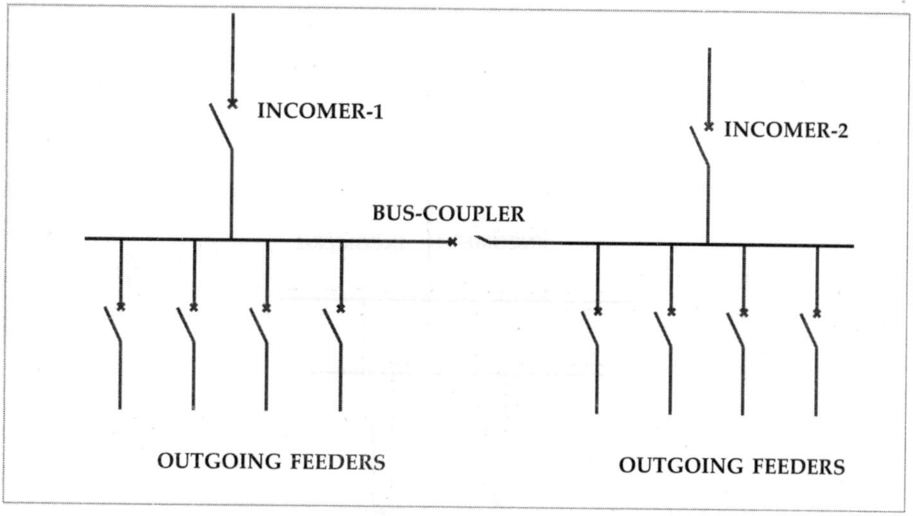

Figure 3.2: Sectionalized bus-bar scheme

Sectionalized ring bus-bar scheme

In sectionalized ring bus-bar system, both the open ends of the bus-bars are joined through another bus-bar (Refer to Fig. 3.3). It has more flexibility and each section of the bus can feed power to another section.

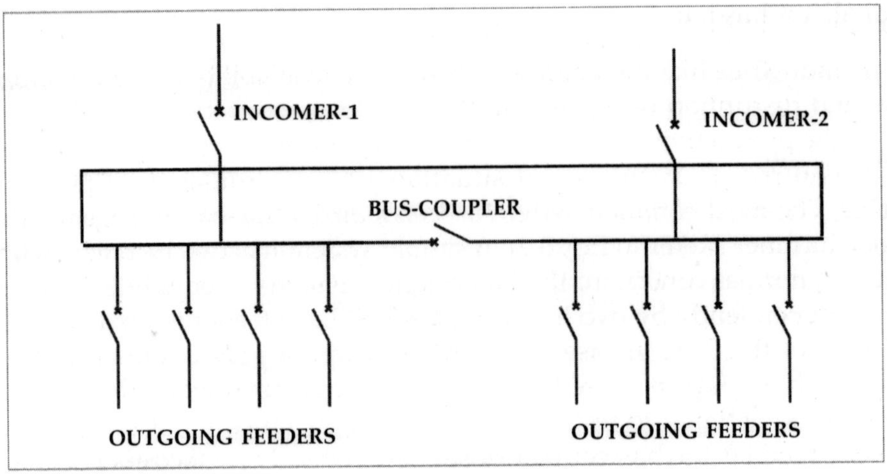

Figure 3.3: Sectionalized ring bus-bar scheme

Duplicate Bus-bar Scheme

In duplicate bus-bar system, two bus-bars are provided (Refer to Fig. 3.4). The bus-bars are supplied with separate sources of power supply. The load is connected to the bus-bars by the on-load changeover switches, so that in case of failure of one source or fault on one bus, the power to the load can be fed through another bus.

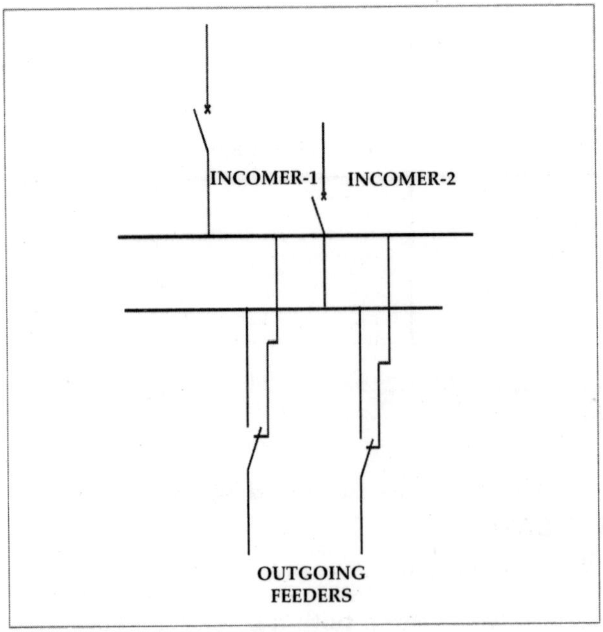

Figure 3.4: Duplicate bus-bar scheme

Sectionalized duplicate bus-bar system

It is a combination of sectionalized bus-bar system and duplicate bus-bar system.

Tie Bus System

Sometimes it is needed to join two PCC through the tie bus. For example, the generator PCC and grid PCC could be joined through two tie bus bars to increase the flexibility of the system (Refer to Fig. 3.5).

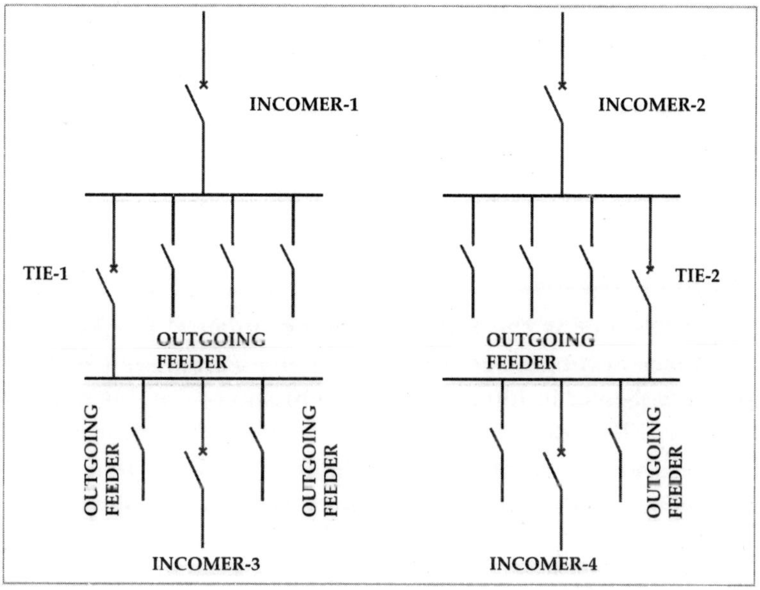

Figure 3.5: Tie bus scheme

3.2 Switching Devices

In power control centre the incomers, bus-couplers and outgoing feeders are connected to circuit breakers. These circuit breakers work as devices for switching on and off the normal power supply and also act as protective devices to isolate the faulty part of the system. In the following section we would briefly discuss the various HT and LT switching devices.

3.2.1 HT circuit breaker

Three types of circuit breakers are being used in substations. The basic difference between these circuit breakers are the type of arc-quenching medium. There are three types of arc-quenching medium.

- Oil
- Vacuum
- Sulphur hexafluoride gas (SF6)

To decide which type of circuit breaker is more suitable for a particular HT power control centre, first of all both technical and non-technical aspects should be analysed. The non-technical aspects for selection of circuit breaker are:

- Economics
- Life
- Availability of spares and service personnel
- Maintenance
- Type of environment
- Safety
- Size
- Frequency of operations

The oldest circuit breaker is the bulk oil circuit breaker. Further advancement in technology took place and bulk oil circuit breaker was replaced by the minimum oil circuit breaker. The bulk and minimum oil circuit breakers are cheaper than the vacuum and sulfur hexafluoride type circuit breakers, but regular maintenance and skilled manpower are required to maintain these circuit breakers. Oil-type circuit breakers are obsolete nowadays, and their manufacturing has also stopped.

The technical aspects are:

- Rated voltage, which will depend on system voltage
- Short-circuit rating
- Normal current rating
- Indoor/Outdoor type

In the following paragraphs, the above-mentioned three types of circuit breakers are discussed briefly.

The basic purpose of providing a circuit breaker is to make and break normal as well as fault current. It is a kind of switch but the difference between a switch and a circuit breaker is that the switch can only make and break the normal current, while a circuit breaker can make and break both normal as well as fault current. The circuit breaker consists of moving and fixed contacts. In closed position, both the contacts press against each other. The breaker is opened through a spring mechanism. The separation of contacts causes an arc due to formation of hot spot on the surface of the contacts.

In oil circuit breaker the oil insulates the live contacts; it provides an insulating barrier. The arc produces hydrogen during arcing period, which helps in recovering the dielectric strength quickly during current-zero period. The contacts separate inside a special enclosure and the blast of hydrogen gas (70%) and mixture of other gases escapes through the vents provided in the enclosure and the arc is forced by the blast into the vents, and fresh cool oil is filled into the enclosure. Thus the arc is extinguished. The oil type circuit breaker occupies more space in comparison to the vacuum and SF6 type of circuit breakers.

In vacuum circuit breaker, the separation of contacts takes place into high vacuum, 10^{-5} torre environment. The current in the arc is carried by the metal ions from the contacts. The metal ions are condensed on the contacts during the current zero and the arc is extinguished.

In SF_6 circuit breaker, SF_6 is an electronegative gas and its molecules rapidly absorb the electrons in the arc path between the contacts and forms negatively charged ions, which are not current carriers. This results in a rapid build-up of insulation strength on current zero.

Both vacuum and SF_6 type circuit breakers have long life and very little maintenance is required.

A comparison is made between the above-mentioned circuit breakers in Table 3.1.

3.2.2 Comparison between different types of circuit breakers

Table 3.1

Characteristics	Bulk/Minimum Oil circuit breaker	Vacuum circuit breaker	SF_6 circuit breaker
Installation	It requires more space in comparison of vacuum and SF_6 type of circuit breaker. The operating mechanism is also bulky and in metropolitan cities where availability of space is one of the constraints, this type of breaker is not suitable.	Requires less space.	Requires less space.
Operation and maintenance	It requires a regular maintenance, and skilled manpower is required for the maintenance. If the	It requires almost negligible maintenance. Circuit breakers with 10,000 operations with little	Similar to vacuum circuit breaker; negligible maintenance is required.

	frequency of operation is more, replacement of oil, checking of wear and tear of contacts etc. will be more frequently required.	maintenance are available nowadays.	
Safety and reliability	Oil is used for extinguishing the arc, and inflammable gases are released during the operation of the circuit breaker. But the design of a circuit breaker is proven and all the aspects of fire and human safety are taken care of. Occurrence of fire due to oil or released gases is rare.	Completely fire safe and risk free.	Completely fire safe and risk free.
Suitability for Industrial application	Before the development of Vacuum and SF_6 circuit breakers, Oil circuit breakers had given a long and trouble-free service to the industry and still in many industries these type of circuit breakers are being successfully used. However, nowadays manufacturing of oil-type circuit has completely stopped.	Presently, vacuum circuit breakers are the most popular ones in the industry. Reason being with the increase in the size of industry, the demand has also increased and consequently, the fault level has increased too. With the advancement in technology, vacuum circuit breakers are fully capable of handling a high level of fault current.	SF6 type of circuit breakers are also popular but these are more in use in high and extra high voltage system.

Coordinated values of rated voltages, short-circuit breaking current and rated normal current are given in Table Xa and Xc in IS 13118: 1991.
The same is reproduced below with the permission of Bureau of Indian Standards

Table Xa: Coordination table of rated values for circuit-breakers

Rated Voltage U (KV)	Rated short circuit breaking current Isc (KA)	Rated Normal Current I_n (A)							
3.6	10	400							
	16		630		1250				
	25				1250	1600		2500	
	40				1250	1600		2500	4000
7.2	8	400							
	12.5	400	630		1250				
	16		630		1250	1600			
	25		630		1250	1600		2500	
	40				1250	1600		2500	4000
12	8	400							
	12.5	400	630		1250				
	16		630		1250	1600			
	25		630		1250	1600		2500	
	40				1250	1600		2500	4000
	50				1250	1600		2500	4000
17.5	8	400	630		1250				
	12.5		630		1250				
	16		630		1250				
	25				1250				
	40				1250	1600		2500	
24	8	400	630		1250				
	12.5		630		1250				
	16		630		1250				
	25					1600		2500	
	40					1600		2500	4000
36	8		630		1250				
	12.5		630		1250	1600			
	16		630		1250	1600			
	25					1600		2500	
	40							2500	4000
52	8			800	1250				
	12.5				1250	1600	2000		
	20								
72.5	12.5			800	1250				
	16			800	1250				
	20				1250	1600	2000		
	31.5					1600	2000		

Table Xc: Coordination table of rated values for circuit-breakers

Rated Voltage U (KV)	Rated short circuit breaking current Isc (KA)	Rated Normal Current I_n (A)					
123	12.5	800	1250		2000		
	20		1250	1600	2000		
	25		1250	1600	2000		
	40			1600			
145	12.5	800	1250		2000		
	20		1250	1600	2000		
	25		1250	1600	2000	3150	
	31.5		1250	1600	2000	3150	
	40			1600	2000	3150	
	50						
170	12.5	800	1250		2000		
	20		1250	1600	2000	3150	
	31.5		1250	1600	2000	3150	
	40			1600	2000	3150	
	50			1600			
					2000		
245	20		1250	1600	2000		
	31.5		1250	1600	2000	3150	
	40			1600	2000	3150	
	50						
					2000		
300	16		1250	1600	2000		
	20		1250	1600	2000		
	31.5		1250	1600	2000	3150	
	50			1600			
					2000		
362	20				2000	3150	
	31.5				2000	3150	
	40				2000	3150	
				1600			
420	20				2000		
	31.5			1600	2000	3150	
	40			1600			
	50			1600			
					2000	3150	
525	40						4000
					2000	3150	
765	40						

3.2.3 Specifications

Before preparing tender enquiry for ordering HT switch-gear, specifications are required to be finalized. The technical requirement are as indicated below:

- Rated voltage (KV)
- Rated insulation level
- Rated lightning impulse withstand voltage (KV)
- Rated switching impulse withstand voltage (KV)
- Rated frequency
- Rated normal current
- Rated short circuit withstand current
- Rated peak withstand current
- Rated duration of short circuit
- Rated short circuit breaking current
- Rated short circuit making current
- Control supply voltage
- Rated operating sequence
- Auxiliary contacts
- Spring charging mechanism, manual or motorized

The circuit breaker is truck mounted for wheeling in and out. It has three positions, service, test and isolation. Safety shutters are provided for keeping the isolation from the live bus.

3.2.4 Low-voltage protective devices

Following low-voltage protective devices are most frequently used in any electrical system.

- Miniature Circuit Breaker (MCB)
- Moulded Case Circuit Breaker (MCCB)
- Air Circuit Breaker (ACB)
- Fuses
- Switch Disconnector fuse unit

Miniature circuit breaker

The miniature circuit breaker performs two functions:

- Protect the system from overload and short circuit.
- Works as a switch during normal condition for switching ON and OFF the circuit.

Construction

The housing is made from a special-grade plastic material. The mechanism inside the housing consists of spring-operated contacts, bimetallic strip for overload protection and electromagnet for short circuit protection. The contacts are made from special silver and copper alloy. The bimetallic strip deflects in response to the current passing through it and moves against the trip bar which releases the trip mechanism. While in case of short-circuit protection, when the current reaches a predetermined level, the solenoid in the magnetic coil directly pulls in the plunger, which separates the contacts and simultaneously releases the trip mechanism.

Specifications

The following specifications are needed whenever procurement of MCBs are done.

- Number of poles
- Characteristic
- Breaking capacity
- Rated voltage
- Rated current
- Frequency
- Enclosure
- Mounting
- Ambient temperature

Characteristics

Normally, MCBs of three type of characteristic 'B', 'C' and 'D' types are available, depending on the maximum and minimum values of magnetic trip. Reference standard is IS: 8828. Refer to Table 3.2.

Table 3.2

Type	Im 1	Im 2	Application
B	3 In	5 In	Normally used in lighting load
C	5 In	10 In	Typical application is in lighting circuit protection, where starting current is high
D	10 In	20 In	Protection of cables and other circuits where starting current is very high

Im1 = Hold point Im2 = Trip point

Selection

The following points shall be taken into account while selecting the MCB for a particular application:

- Application requirement and utilization category.
- Characteristic of the load to be protected by MCB and accordingly, the MCB of B, C or D type of characteristic should be selected.
- Fault current to be interrupted.
- Number of poles, system voltage and load current.
- Coordination with upstream protective device like MCCB, Fuse, MCB etc.

The other important factors while selecting the MCB are:

- Low power loss.
- Current limiting feature.

Low power loss

It should conform to low power loss as stipulated in IEC898/95 and IS 8828/96. It contributes significantly in energy saving.

Current limiting

The MCB shall have the feature of current limiting and it should trip before the full prospective fault current is developed. The let through (i^2t) energy depends on the current and the time for which it passes. The current can be reduced by reducing the tripping time, and subsequently, the damage to the load is minimized. The current limiting is possible by passing the current in the MCB in such a way that a repulse magnetic field is produced, which forces the contacts to separate even before the operation of the tripping mechanism.

Moulded case circuit breaker

The construction and operating system of MCCB is the same as of MCB, but it has certainly more features than the MCB. A comparison between MCCB and MCB has been done in Table 3.3.

Table: 3.3

Comparison between MCCB and MCB

Sl. No.	Features	MCCB	MCB
1.	Ratings		
	• Current	The current rating is more than MCB, Presently, MCCB of 1600 Amp current rating is available.	MCB of current rating Up to 63 A MCB is available.
	• Voltage	500V	500V
	• Short Circuit Current	More than MCB, upto 50 KA rating MCCB is available.	Up to 10 KA is available.
2.	Settings		
	• Magnetic	Variable, can be manually set, microprocessor based is also available.	Fixed, Factory set.
	• Thermal	Variable, can be manually set, microprocessor based is also available.	Fixed, Factory set.
3.	Shunt tripping	Possible	Not possible
4.	Remote Operation	Possible	Not possible
5.	Under voltage release	Can be fitted	Cannot be fitted
6.	Auxiliary Contacts	Available	Not available
7.	Earth leakage	Separate earth leakage relay can be fitted	Cannot be fitted. But a combination of MCB and ELCB is available
8.	Rotating handle and motor spring charging for remote closing operation.	Can be fitted	Cannot be fitted

The MCCB is the main competitor of air circuit breaker. MCCB occupies less space than the air circuit breaker and is more useful where there is a space constraint.

Specifications

- Number of poles
- Nominal current

- Rated operating voltage
- Rated insulation voltage
- Rated impulse withstand voltage
- Category of use
- Breaking capacity
- Making capacity
- Rated short time withstand capacity
- Thermal magnetic releases
- Electronic releases
- Earth-leakage module
- Remote control (Shunt trip)
- Fixed type/Draw-out type
- Rotary handle
- Auxiliary contact
- Under voltage release
- Motor for spring charging

Selection

The following points shall be taken into account while selecting the MCCB for a particular application.

- Application requirement, utilization category.
- Short circuit withstand capacity.
- Fault current to be interrupted.
- Number of poles, system voltage and load current.
- Coordination with upstream protective device like MCCB, Fuse, MCB etc
- Other accessories like rotary handle, remote operation, earth leakage module, auxiliary contacts etc.

Air circuit breaker

Air circuit breaker is a traditional device in low-tension electrical system. It can close, open a healthy system and also trip whenever a fault takes place in the system. It is a very rugged device and quick-make-and-break type. It has a very high breaking capacity of 50 KA. However, circuit breakers of breaking capacity of 120 KA are also available. They need very little maintenance. The arc is extinguished by lengthening and splitting the arc. While preparing tender enquiry for procurement of air circuit breaker, the following specifications are required to be provided to the vendors.

Specifications

- Rated current
- Number of poles
- Insulation voltage
- Breaking capacity
- Making capacity
- Mounting
- Tripping devices (like overcurrent and earth fault releases), shunt trip
- Accessories like NO/NC contacts
- Control supply voltage
- Spring charging arrangement i.e. motorized or manual
- Auxiliary contact
- Draw out or fixed type

Selection

While selecting air circuit breaker, the following points should be considered:

- Current rating
- Mounting: Fixed, draw-out type
- Number of poles
- Short circuit capacity
- Accessories: Safety shutter, position switch
- Type of spring charging system: Manual/motorised
- Releases: Short-circuit, over-current, ground-fault, solid-state/microprocessor type
- Tripping devices: Shunt trip, no volt and under-voltage trip
- Breaking and making capacity
- Auxiliary contacts

Coordinated values of rated current and short circuit current of LT air circuit breaker

Coordinated values of rated current and short circuit current are given in Table 3.4.

Table 3.4

Rated short circuit breaking current Isc (KA)			Rated Normal Current In (A)									
25		630										
36.6	400	630	800	1000								
38			800	1000	1250	1600						
50			800	1000	1250	1600	2000					
55							2000					
65			800	1000	1250	1600		2500	3200	4000		
70							2000					
75								2000	3200	4000		
85											5000	6300
100									3200	4000	5000	6300

The following two types of short circuit breaking capacity are specified in the circuit breaker specifications:

- Rated ultimate short circuit breaking capacity
- Rated service short circuit capacity

The rated ultimate short circuit breaking capacity is the rms value of the prospective current assigned to the circuit breaker by the manufacturer without taking into account the short time withstand capability of the circuit breaker. It is indicated as Icu.

The rated service short circuit capability takes into account the short time withstand capability of the circuit breaker. It is indicated as Ics. It is expressed as a value of prospective breaking current corresponding to the one of the specified percentage of the rated ultimate short circuit breaking capacity.

Table 3.5

Standard ratios between Ics and Icu

Utilization category A (% of Icu)	Utilization category B (% of Icu)
25	-
50	50
75	75
100	100

Table 3.6

Standard relationship between short circuit making and breaking capacity and related power factor.

Short circuit breaking capacity (KA) (r.m.s.)	Power factor	Minimum value required for N N= short circuit making capacity/ short circuit breaking capacity
4.5<I<=6	0.7	1.5
6<I<=10	0.5	1.7
10<I<=20	0.3	2.0
20<I<=50	0.25	2.1
50<I	0.2	2.2

Preferred values of the short time delays:

(sec.)

0.05

0.1

0.25

0.5

1.0

Table 3.7

Minimum values of rated short time withstand current:

Rated Current In (A)	Rate short time withstand current Icw(KA)
In<= 2500	12 In or 5KA, whichever is greater
In>2500	30 KA

Table 3.8

Utilization category:

Utilization category	Application with respect to selectivity
A	Circuit breakers are not specifically intended for selectivity under short circuit conditions with respect to other short circuit protective devices in series of load side, i.e. without an intentional short-time delay provided for selectivity under short circuit conditions, and therefore, without a short time withstand current rating.
B	Circuit breakers specifically intended for selectivity under short circuit conditions with respect to other

short circuit protective devices in series on the load side, i.e. with an intentional short time delay (which may be adjustable), provided for selectivity under short circuit conditions. Such circuit breakers have a short time with stand current rating.

Switch disconnector fuse unit

The advantage of switch disconnector fuse unit is that the fuses are isolated at both ends and can be replaced easily. It is a quick-make-and-break type and front operated.

Specifications

The following are required to be specified at the time of procurement of the switch disconnector fuse unit:

- Number of poles
- Neutral terminal
- Voltage rating
- Insulation withstand voltage
- Frequency
- Rated current
- Utilization category
- Rated making current
- Rated breaking capacity
- Rated fused short circuit capacity

Comparison between air circuit breaker, fuses, miniature circuit breaker and moulded case circuit breaker.

A comparison between air circuit breaker, moulded case circuit breaker, miniature circuit breaker and HRC fuse is given in Table 3.9.

Table 3.9

Sl. No.	Characteristic	Air circuit breaker	Moulded case circuit breaker	Miniature circuit breaker	HRC Fuse
1.	Short circuit level	High upto 120 KA	High upto100 KA	Low upto10 KA	High upto 100 KA
2.	Protective Devices	Microprocessor based S/C and O/C, devices with variable	Microprocessor based S/C and O/ C,Electro-magnetic S/C and thermal	Fixed, factory set thermal and magnetic devices	Not available. Only fusing element inside

		settings. Electro-magnetic S/C and thermal for O/L protection devices with variable settings	for O/L		the quartz granule
3.	Operation	High Speed operating time of 30ms including 10 ms of arcing time	High speed operation	High speed operation	Quick clearing of fault
4.	Maintenance	Low maintenance	No Maintenance	No maintenance	Required to be replaced after fault clearing
5.	Cost	Higher than MCCB	Lower than ACB	Low Cost	Low Cost
6.	Fault Current Cut Off	Fault current is limited and interrupted before it reaches its peak value	Fault current is limited and interrupted before it reaches its peak value	Fault current is limited and interrupted before it reaches its peak value	Fault current is cut off much earlier than reaching its peak value
7.	Earth fault over-current and short circuit clearing	Available through releases and adjustable	Available	Operates on short circuiting and overloading. Settings are fixed and factory set	Operates on short circuiting
8.	Remote Operation	Can be done	Can be done	Not possible	Not possible
9.	Arc Quenching Medium	Air	Air	Air	Quartz granule
10.	Current range	600 A to 6300A	16 A to 1600 A	0.4 A to 63 A	2 A to 630 A
11.	Installation	Draw-out, Fixed	Draw-out, Fixed	Fixed	Fixed
12	Coordination	High Degree of Coordination	High Degree of Coordination	Possible	Possible

3.3 General Requirement of Power Control Centre (PCC)/Motor Control Centre Panel (MCC)

Free standing, modular type

Modules may be fixed type or draw-out type with three positions like service, test and isolation. Draw-out type modular construction has some advantages over fixed type. The same modules can be interchanged. For example, if some fault takes place in one of the modules, it can immediately be replaced by the spare modules of same rating. Secondly, it can be tested in test position without feeding the power to the load. Maintenance is more easy since repairing can be done after taking out the complete module.

Cable entry

It may be from the top as also from the bottom. If cables are installed in overhead cable trays and the PCC or MCC location is such that the cable trenches cannot be provided, then cable entry from top of the panel is envisaged. Otherwise in normal case, cable trenches are installed and cable entry from the bottom of the panel is kept. The panel must have sufficient space to accommodate and terminate cables of the specified sizes. In some cases bus-duct is provided as incomer, and it should be clearly specified in the specifications. In case of bus-duct, top entry is preferred.

Extension

The arrangement should be such that the panel could be extended from both sides as per the future requirement.

Enclosure class

Enclosure class like IP-45,IP-51 etc. is specified as per the requirement.

Environmental conditions

It is important to specify the reference ambient temperature, maximum and minimum ambient temperature, height from sea level, humidity etc. The performance of equipment depends very much on the environmental conditions.

Painting

The treatment of the sheet metal like seven tank treatment process and the shade of paint, type of paint like spray paint, power coating etc. should be specified.

Thickness of metal sheet

Normally, thickness of 14 SWG for compartments and 16 SWG for doors etc. is required.

Bus-bars

Voltage, current, short circuit current (r.m.s. value) for 1 sec., peak short circuit, control voltage, etc. shall be specified. Temperature rise limit should be specified. Please refer to Chapter 5 for more details.

Other general requirements are:

1. Horizontal installation of main bus-bars
2. Electrical grade aluminium for main bus-bars
3. Copper auxiliary bus-bar for control circuit
4. Heat shrinkable sleeves for bus-bars with phase identification
5. Bus-bar supports with non-hygroscopic insulators
6. Earth bar at bottom for entire length of the panel
7. Shrouds at joints and tap-off points of bus-bars

3.4 Protection and Measuring Schemes

The type of faults which are encountered in an electrical system generally are overload, short circuiting due to failure of insulation or due to some external element, and earth fault. The protection scheme consists of instrument transformers i.e. current and voltage transformers, relays like over-current, earth fault, differential relays, reverse power relays, master trip relays, auxiliary relays etc. Now the question is which type of relay should be provided to protect a system. A brief answer is that it depends on the equipment type, rating, backup protection etc. For example, over-current and earth fault protection is provided for feeder protection in a system of small capacity. Single phasing protection is provided in addition to over-current and short circuit in motors. In large motors negative phase sequence protection is also provided. In case of alternators reverse power relays (if operating in parallel) are provided in addition to over-current and earth fault relays. At the same time, if the capacity is large, then differential protection is also provided. The relays have time and current setting for its operation and this time and current setting is used for coordination of relays, so that the relay of the faulty part should operate first and only faulty section is isolated leaving the healthy system untouched. If the local relay fails to operate then only the back-up protection should work.

Meters are provided to measure the various electrical parameters like voltage, current, power factor, energy etc. The measurement scheme consists of current and voltage transformer and sometimes transducers too. The electrical systems in industrial and commercial establishments have become more complex and with the advancement in technology, reliability has also increased. The cost of outages due to failure in electrical system particularly in a process industry has also increased. Many industrial plants particularly process plants have their own captive power plants. Besides, diesel generators are also provided as emergency power source and the supply from diesel generators are fed to limited bus-bars, which are sometimes called critical bus-bars to have a safe shut-down of the plant in case of failure of power supply.

Indication and annunciation

Alarm and indicators for each type of fault and subsequently tripping of feeder shall be provided. Other indicators like control supply healthy, three phases (Red, Yellow, Blue) shall also be provided.

Miscellaneous

Provision of anti condensation heaters with thermostat inside the PCC/MCC panel, lights and power sockets shall be made.
The above mentioned points are elaborated in the following example.

Example 3.1

In Chapter 1, a detailed calculation was done for LT load of a steam generation plant of a process industry, and for a total load of 887 KW two transformers of 1000 KVA rating were chosen. The HT side of transformers are connected to the 11 KV bus-bar system. The following steps are taken to proceed further for development of complete protection, measurement and bus-bar scheme.

Step 1

Before development of scheme the following points are to be considered

- Indoor or outdoor type substation.
- Fault current capacity of the system.
- Bus-bar arrangement.
- Numbers of incoming and outgoing feeders.
- Protection scheme.
- Electrical parameters metering scheme.
- Type of circuit breaker.
- Safety requirements.
- Extension of panel in future.
- Requirements of state electricity board.

The substation which we have considered is the indoor type and receives power supply from two separate buses of 11 KV switchgear. The outgoing feeders on these buses feed two transformers of 1000 KVA. It consists of draw-out type truck mounted vacuum circuit breakers with over-current and earth fault protection relays connected to the current transformers. The reason for choosing vacuum circuit breaker rather than oil circuit breaker is that though the oil circuit breakers are cheaper and robust, they have other disadvantages like oil leakages, routine maintenance etc. Further, due to low demand in the industrial and commercial sector, the manufacturing of oil-type circuit breakers have stopped though these type of breakers are serving well in old industry, but these are being removed in a phased manner due to non availability of spares.

One most important point has been left out: What should be the fault current carrying capacity of the system? The answer is that it should be as per the source feeding the system. Now, if the source is state electricity board feeder, the information is to be collected from state electricity board like what the capacity of the transformer is from which the power will be supplied, and how far it is located. Do they have any plan to replace it with higher capacity of transformer in near future? The information of fault current capacity of the electricity board substation feeding the HT supply shall be obtained.

In this example, power supply will be received from 11KV system of the plant and the fault capacity of the 11 KV system is 500 MVA. The bus-bars of 11 KV system are arranged in a ring-type arrangement. The maximum short circuit current for one second is calculated in step (d) (Example 3.1). The short circuit current comes out to be 23.96 KA.

The rating of circuit breaker should be according to that fault capacity. Now, refer to Table Xa of this chapter, the circuit breaker available for this fault level will have the following parameters:

System Voltage: 12 KV

Short circuit capacity: 40 KA for 1 sec.

Normal rated current: 1250 A

Step 2

Next step is to give a thought to the protection and measurement schemes for above 11 KV feeders.

A simple scheme with two over-current and one residually connected earth fault relay with master trip relay will be adequate. The present trend is to provide microprocessor-based communicable relays, these relays are compact i.e. one unit contains over-current as well as earth fault relays and consists of many additional features. The auxiliary relays like Buchloz relay, winding temperature alarm and trip relay and oil temperature alarm and trip relay shall be provided for protection of oil-type transformer. If the transformer is of dry type, the oil temperature relays are not required. The dry type transformers have embedded RTD (Resistance Temperature Detector) for winding temperature alarm and trip system.

As far as measuring devices are concerned, the incomers and outgoing feeders should have multi-parameter microprocessor-based measuring devices. The most important thing with the measuring system is the monitoring of the power consumption of different feeders' load. The multi-parameter measuring devices with communicable facility with the computers are very useful tool in electrical energy monitoring.

The design of the system shall be such that no live part shall be exposed to humans. It shall also have provision for extension of PCC/MCC in future if required.

For electrical system of steam generation plant, current transformers are required for measurement and protection while potential transformer is required for measurement only. The standard practice is to provide $110/\sqrt{3}$ star connected potential transformers. The current transformers for measurement shall be provided according to the rating of the transformer. The rated current of the 1000 KVA transformer will be 52.85 A at 11 KV side and considering 10% overload capacity the CT ratio of 60/1 will be adequate. The class of CT will be "Class 0.5".

The protection CT depends on the setting of the relay connected with it and relay setting depends on time and current grading for proper coordination of the system. The protection class is 5P10. For the time being, we take the Protection CT of current ratio of 60/1. Both CT and PT are of cast resin type. The other important factor in selection of current transformer is the VA burden. VA burden depends on the impedance of the relay, connected wiring and VA burden should be chosen in such a way that it could drive sufficient current in the secondary for proper operation of the relay. With provision of modern electronic relays CTs of less VA burden can be safely used.

Step 3

The HT system in our example is defined, and we could proceed to draw a single-line diagram of HT system. Refer to Fig. 3.6.

Having developed the single line diagram, technical specifications of HT power control centre is developed. Sample specifications are given below:

Circuit breaker

- Type: Indoor
- Arc quenching medium: Vacuum
- Rated Voltage: 12 KV
- Power Frequency withstand Voltage: 28 KV
- Lightning Impulse Voltage: 75 KV
- Rated current: 1600 A

- Short circuit making current: 110 KA,1 sec
- Short circuit breaking current: 40 KA,1 sec
- Short circuit withstand capacity (Symmetrical): 40 KA,1 sec
- Spring charging mechanism: Motorized, 230 V AC
- Closing and tripping voltage: 110 VDC
- Mounting: Truck mounted with three position (service, test and isolated)
- Safety features: Safety shutters shall be provided

HT Panel

- Bus-bar location: Top
- Type: Single front, free standing, cubicle type
- Fault level: 40 KA rms, 1 sec
- Rise in temperature: 90^0C above ambient
- Earth bar: Copper
- Bus-bar: Copper, sleeved and joints shrouded
- Bus Rating: 1600 A
- Control wiring: 2.5 sq mm, 1.1/0.66 KV, PVC insulated
- Enclosure: IP55
- Thickness of sheet: 2 MM
- Cable entry: bottom

Feeders

- Rating: 11KV
- Protection CT: Cast resin, 5P10,60/1, 15 VA
- Metering CT: Cast resin type, Class 0.5, 60/1, 15 VA
- Metering PT: Cast resin, 11 KV/√110 V, 15 VA
- Over current and earth fault relay: Microprocessor based with facility for communication with computer through RS 232/ RS 485 interfacing port.
- Annunciation: Over-current, earth fault, winding temperature trip, DC supply healthy
- Measuring devices: Multi parameter, microprocessor based with facility for communication with a PC through RS 232/ RS 485 interfacing port.

Step 4

LT System

Since two transformers are provided, the LT system shall have two incomers and one bus-coupler. As mentioned earlier, in an industrial system, particularly in a process plant, reliability of power supply is very important, and it is a normal practice that each feeder should be capable to take 100% load so that power supply could not be interrupted in case of fault on the other incomer. Auto changeover scheme is also provided so that on failure of power supply of either incoming feeder, the bus coupler closes automatically to feed the bus whose incomer has tripped provided there is no fault on that bus-bar.

Besides the above two incomers, which are fed by the 1000 KVA transformer, we have another incomer called emergency incomer for feeding power supply to emergency load. The emergency incomer is connected to a separate bus called emergency bus. A circuit breaker is provided between normal and emergency bus as a bus-coupler. In normal case when the normal power supply is available, emergency bus is fed by closing the bus coupler and the circuit breaker of incomer of emergency bus remains

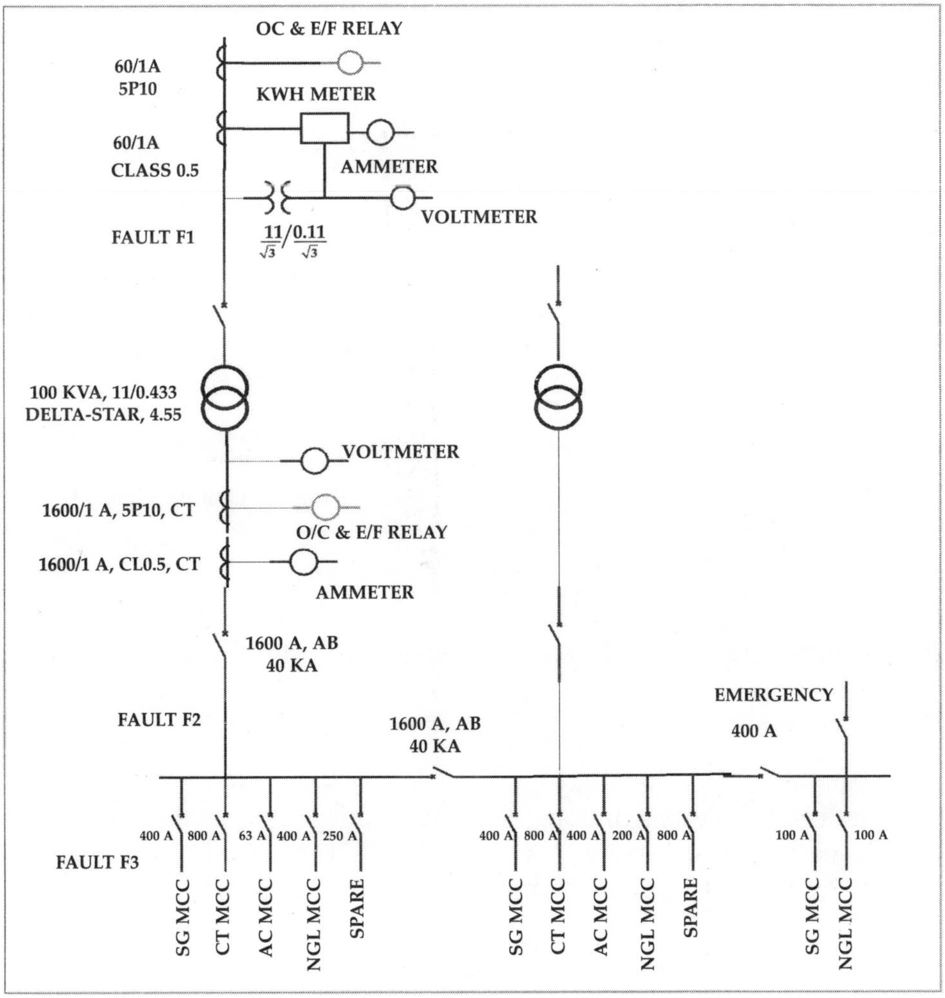

Figure 3.6: Single line diagram of electrical system

in service position, but in off condition. On failure of power supply, the emergency bus-coupler trips so that it is disconnected with the normal bus and the emergency incomer closes. The power supply to emergency bus is fed through an emergency source like a DG set. Sometimes the emergency system has a separate emergency panel instead of providing an emergency bus in the same (normal) panel. This type of emergency panel has two incomers, and one of the incomer is connected with normal power supply source while another is connected with emergency source. Only one incomer remains ON at a time.

To accomplish the auto changeover scheme between emergency bus-coupler and emergency incomer, a control circuit having contactors, timers, interlocks etc. is provided. Care should be taken by providing electrical and mechanical interlocks so that two sources don't get paralleled.

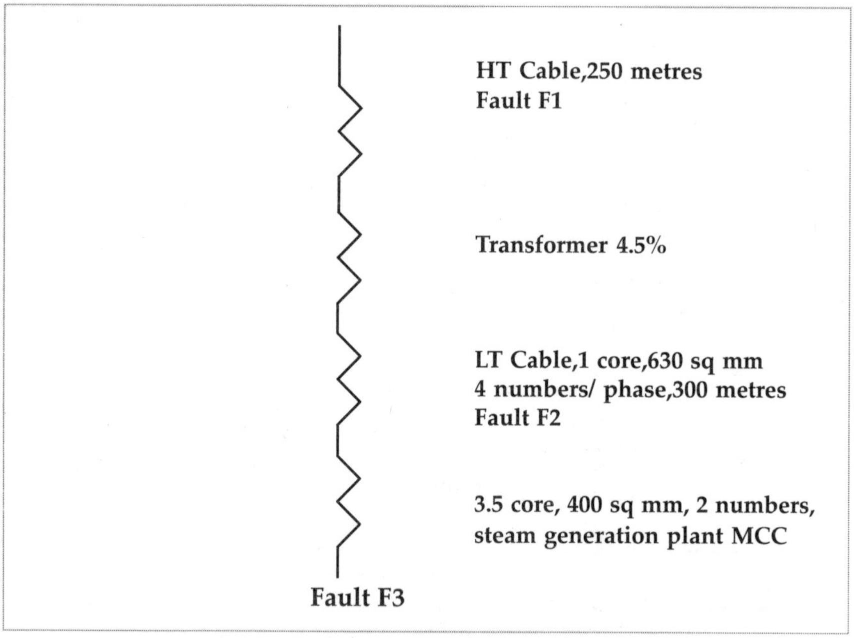

Figure 3.7: Reactance diagram for symmetrical fault current calculation

Step 5

We have planned the incomers and bus-bars and left-out portions are the cables and outgoing feeders. The details of the various loads are given in Table 1.14, Chapter 1. These loads are further sub-divided into various MCC (motor Control Centre) Loads as per Table 3.10 to Table 3.14 of this chapter.

Step 6

The most important step is calculation of fault current at different locations. However, before calculation of fault current, it is important to draw a single line diagram. Refer to Fig. 3.6.

First of all, we will calculate fault current at LT bus-bar so that the fault rating of the bus, circuit breaker could be finalized. After this fault at load end will be calculated. Relay coordination will also be highlighted.

Example 3.2

Calculation of fault at F1 and F2 refer to Fig. 3.6

Step a

Prepare a reactance diagram of the power system. Refer to Fig. 3.7.

Step b

Select the base KVA. We have selected 1000 KVA as base.

Step c

Various reactances are converted in per unit value on the chosen base.

- 11 KV system, Fault MVA: 500 MVA :

  1000 KVA/500000 KVA = 0.002 per unit
- 11 KV cable: 0.092 ohm/KM

  length of cable = 250 mts

  Reactance = .092 x 250/1000 = 0.023

  = Ohm x Base KVA/(1000 x KV^2)

  = 0.023 x 1000/(1000 x 11^2)

  = 0.00019 per unit
- Transformer % impedance = 4.5% = 4.5/100 = 0.045

  At base KVA = per unit x base KVA/KVA rating

  = 0.045 x 1000/1000 = 0.045 per unit
- LT Cable: 4 Numbers of 630 sq mm, single core, per phase and of approximately 300 metres length

  Reactance 0.077 ohm/km

  = 0.077 x 300/1000 = 0.0231 ohm

  since four cables are in parallel

  = 0.0231/4 = 0.00577 ohm

Per unit value = $0.00577 \times 1000/(1000 \times 0.415^2)$

= 0.0335 per unit

Step d

Fault at F1

Total per unit reacance 0.002 + 0.00019 = 0.00219
Symmetrical short circuit current = 1000 (Base KVA)/(1.732 x 11 x 0.00219)
= 23.96 KA

Step e

Fault at F2

Total per unit reactance = 0.002 + 0.00019 + 0.045 + 0.0335 = 0.0807 (Refer reactance diagram in Fig. 3.7)

Symmetrical short circuit current for fault F2 will be

= 1000 (Base KVA)/(1.732 x 0.415 x 0.0807)

= 17241.38 amp

= 17.24 KA

17.24 KA, (rms value) is the maximum current, which can pass through the LT circuit breaker and bus-bars at the time of symmetrical fault on LT bus-bars. Therefore, the rating of the circuit breaker and bus-bars should be such that it can withstand the 17.24 KA for at least 1 sec.

In addition, the above bus-bar and circuit breaker should be capable of taking full load plus 10% overload of the transformer capacity which comes out to be 1530.37 amp. Therefore the bus-bar should be rated for near that value and the standard current rating for bus-bar is 1600 amp. Therefore, bus-bars of appropriate size that can take 1600 amp current should be chosen. Refer to Chapter 5 for calculation of bus-bar size.

Step 7

Now we look at the load end. Table 1.14, Chapter 1, gives the total load of the steam generation plant. This load is being fed by different motor control centres and it is segregated in the following Table 3.10 to Table 3.14.

Table 3.10

Steam Generation Plant MCC

Description	Qty.	Voltage	Rating (KW)	Diversity factor	Load factor	Total Load (KW)
Makeup water pump motor	03	415V	90	1.5	80%	144
Air compressor motor	03	415V	115	1.5	70%	161
Air heater	02	415V	30	1	90%	54
Condensate Extraction pump	04	415V	22	2	80%	35.2
Lighting, UPS, Miscellaneous load	-	415V	9	1.8	80%	4
Auxiliary oil pump	02	415V	55	2.5	70%	15.4
Total						**413.6**

Table 3.11

Emergency Load

Description	Qty.	Voltage	Rating (KW)	Diversity factor	Load factor	Total
Makeup water pump motor	01	415V	90	1.5	80%	48
Condensate Extraction pump	04	415V	22	2	80%	35.2
Lighting, UPS, Miscellaneous	-	415V	6	1.8	80%	2.67
Total						**85.86**

Table 3.12

Cooling Tower MCC

Description	Qty.	Voltage	Rating (KW)	Diversity factor	Load factor	Total
Auxiliary water pump	02	415V	75	1.5	80%	80
CT fan motor	04	415V	45	1	60%	108
Lighting	-	415V	2	1.8	80%	0.88
Total						**188.88**

Table 3.13

Air-conditioning MCC

Description	Qty.	Voltage	Rating (KW)	Diversity factor	Load factor	Total
Chiller compressor	03	415V	55	1.5	80%	88
Air handling unit	03	415V	45	1.5	80%	72
Cooling tower, & Miscellaneous other load	06	415	17	2	80%	40.8
Lighting	-	415V	1	1.8	80%	0.44
Total						**201.24**

Table 3.14

NGL (Natural Gas Liquified)

Description	Qty	Voltage	Rating (KW)	Diversity factor	Load factor	Total
NGL firing pump	06	415V	55	3	70%	77
NGL unloading pump	02	415V	7.5	1.5	50%	5
Lighting	-	415V	3	1.8	80%	1.33
Total						**83.33**

As per the philosophy of the electrical system for a process plant, there should be two incomers and one bus-coupler and each incomer should be capable of taking 100% load in case of failure of one incomer. Auto changeover system should be provided. Besides this, spare feeders should be provided for each MCC.

Outgoing feeders shall be as below:

- Steam generation plant MCC: Three feeders of 800 amp. Air circuit breaker
- Cooling tower: Three feeders of 400 amp. Air circuit breaker
- Air-conditioning system: Three feeders of 400 amp. Air circuit breakers
- NGL MCC: Three feeders of 200 amp. MCCB
- One spare feeder for each MCC incomer

Step 8

The last step is to calculate the symmetrical short circuit current at load end. The procedure is similar to the one mentioned above. For illustration we take MCC for steam generation plant. Two numbers 3.5 core and 400 sq mm are connected in parallel for a load of 413.6 KW. The approximate distance is 400 metres.

Reactance of cable 0.086 ohm/ km

Reactance for two cables = 0.086 x 400/(1000 x 2) = 0.0172 ohm

Per unit value = 0.0172 x 1000/(1000 x 0.415^2)

= 0.0998 per unit

We have already calculated the per unit reactance up to LT PCC, which is 0.0807, therefore, the total per unit reactance at load end will be 0.0807+0.0998 = 0.1805

The symmetrical fault current will be

1000/(1.732x0.415 x 0.1805) = 7707.73 amp = 7.71 KA

Example 3.1

We have already discussed the protection system in brief in Section 3.4. The concept can be better understood by the following example. The fault currents at various levels have already been calculated and we have the load details and the single-line diagram. To simplify further, it is better to tabulate these details in Table 3.15.

Table 3.15

	Rated current (A)	Short circuit current (KA)	CT Ratio	Overload protection	Earth fault protection
11 KV	57.7 (Including 10% O/C capacity of transformer)	23.97	60/1	IDMT relay (3 sec) in two phases with high set element.	IDMT residually connected 3 sec
415 V Power control centre	1530 (Including 10% O/C capacity of transformer)	17.24	1600/1	IDMT relay (3 sec) in two Phase	IDMT residually connected 3 sec
Maximum current when largest motor of 115 KW is started with a base load of 890 KW	2314 Amp. at LT side and 87.3 Amp. at HT side				
Steam generation plant I/C	605.7	7.75	800/1	IDMT relay in two phases 3 sec	IDMT residually connected 3 sec

Maximum current when largest motor of 115 KW is started with a base load of 360 KW	1537

Current setting

LT side load end.

Current setting : 100%

Time multiplier : 0.3

Normally, the largest motor takes 3 seconds in picking up the full speed. The relay will take 3 seconds to initiate trip command at time multiple setting of 0.3 (Multiple of plug setting current 1537/800 = 1.92). It is safe enough. At short circuit the multiple of plug setting will be 7750/800 = 9.68 and relay will take 0.92 second to initiate the tripping command. The over current protection relay is inverse definite minimum time relay of 3 seconds.

PCC side relay settings

Current setting: 100%
Time multiplier: 0.3

At short circuit the multiple of plug setting will be 17240/1600 = 10.7 and it will take 0.88 second to initiate the tripping command.

With this setting, if the fault takes place on the load end i.e. on MCC bus of steam generation plant then it will take 1.3 second (Multiple of Plug setting current 7750/1600=4.84, time multiplier 0.3), so the margin is 1.3 – 0.92 = 0.38 second, which is safe enough.

HT side

Current setting: 100%

Time multiplier: 0.6

The transformer should not trip below its rated current plus 10% overload capacity. Therefore, the setting should be 100% i.e. Tap 3 of (0.5 –2.0) relay. Current at HT side if short circuit takes place at LT side 17240 x 0.415/11 = 650 A. Multiples of plug setting current 650/60 = 10.84. Tripping time 1.7 second. Time margin 1.7-1.3 = 0.40 secondwill be alright. If the fault takes place at HT bus, the multiple of plug setting will be 23970/60 = 399.5 and the minimum time will be 1.2 second, which is large enough. Therefore, instantaneous element for short circuit protection is provided.

It is also required to plot the curves on a same current scale. Normally, current at low voltage side is taken as base current for plotting the curves. Such relay coordination curve for the above example has been drawn in Fig. 3.8.

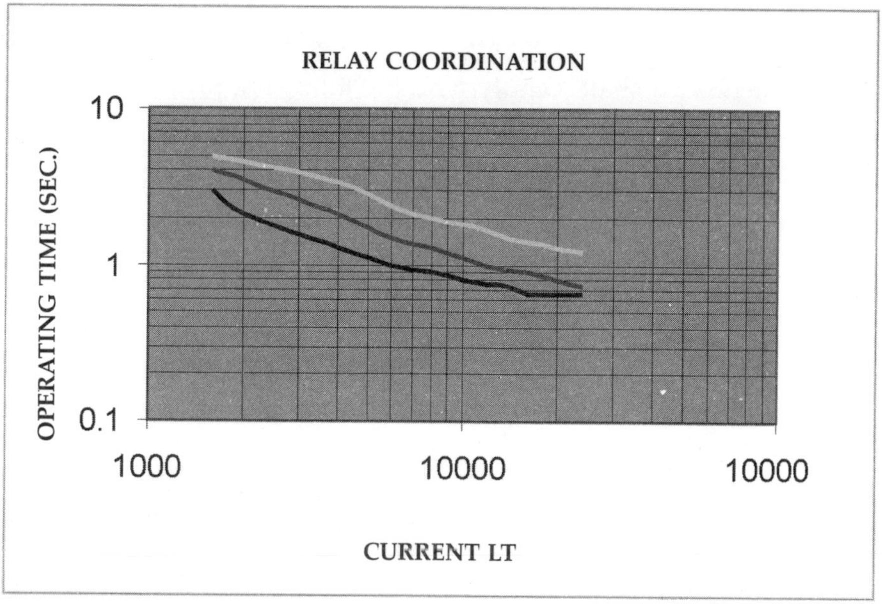

Relay at HT Feeder
Relay at LT I/C
Relay at LT O/G

Figure 3.8: Relay coordination curves

3.5 Installation

The PCC/MCC should be installed as per the standard practice, Indian Electricity Rules etc. Steel channels are provided across the cable trenches on which PCC/MCC is installed. It is bolted or welded with the channel. Normally the MCC/PCC is supplied in sections for ease of transportation. These sections are joined at the sides together at the time of installation. The room shall be properly ventilated. It shall be properly lit and sufficient space shall be provided for maintenance of the panel.

As per the Indian Electricity Rules 51, the following provisions shall be observed where energy at medium, high or extra high voltage is supplied, converted, transformed or used:

1. a) All conductors (other than those of overhead lines) shall be completely enclosed in mechanically strong metal casing or metallic covering, which is electrically and mechanically continuous and adequately protected against mechanical

damage unless the said conductors are accessible only to an authorized person or are installed and protected to the satisfaction of inspector.

b) All metal works, enclosing, supporting or associated with the installation, other than that designed to serve as a conductor, shall be connected with a grounding system as per standards laid down in the Indian Standards in this regard.

c) Every switchboard shall comply with the following provisions:

- A clear space of not less than 1 metre in width shall be provided in front of switchboard.
- If there are any attachments or bare connections at the back of the switchboard, the space (if any) behind the switchboard shall be either less than 20 cm or more than 75 cm in width, measured from the farthest outstanding part of any attachment or conductor.
- If the space behind the switch board exceeds 75 cm in width, there shall be a passage way from either end of the switchboard clear to a height of 1.8 metres.

3.6 Testing

All the routine tests as specified in the Indian standards or other relevant standards should be carried out at site. Some of the tests are insulation test, which is carried out with the help of insulation tester. High voltage tests are carried out by applying a high voltage between the phases and earth for one minute. All the control schemes, interlocking, auto changeover scheme and tripping schemes, should be tested. The protective relays should be tested by applying the current in the relay coils. The polarity, current and voltage ratio of the current and voltage transformer should be tested.

Sometimes it is required to carry out the heat run test on the bus-bars to establish the capacity of the bus-bar. This test is done at the manufacturer's place because large current i.e. equal to the current rating of bus-bars are required to be injected in the bus-bar for a specified period, and generally such facility is not available at site. This is not a routine test and is carried out only when prior agreement is made with the vendor.

3.7 Future Trends in the Power Control Centre and the Motor Control Centre

What are we looking for in a future incomer of a PCC or MCC? In earlier PCC/MCC, we find air circuit breaker as incomer and also switch fuse unit in PCC of low rating. These devices used to be very bulky and occupied a lot of space. The present trend is "FUSELESS" PCC. Sometimes the FUSELESS term is misunderstood as PCC without any protection device. We come back to our original question. It is well known that incomer should have the following properties:

- It should be capable of withstanding the short-circuit current for a certain duration, normally one second.
- It should be capable to clear the fault current safely.

In time-graded protection system, which is commonly used, the circuit breaker that is uppermost is set in such a way that its tripping time is maximum. If the fault takes place just near that circuit breaker, it takes a long time to trip for the above mentioned reason if only time-graded system is provided, but in all standard protection since relay has inverse time characteristic and therefore if current is more, it will take less time to operate. If fault takes place at load end and two or three circuit breakers are in series, then extreme end breaker at the upstream side is set to take more time to trip to have a proper discrimination. Now the trend is to adopt the concept of zone selective system. In this system, the part of installation between two circuit breakers which are connected in series is called one zone. Every downstream circuit breaker is connected with the upstream circuit breaker through a communication cable. Every downstream circuit breaker sends a signal to upstream circuit breaker after detecting a fault. If the fault takes place near a circuit breaker and it has not received any communication from its downstream circuit breaker, it should trip immediately without waiting for any time delay. However, if fault takes place near any downstream circuit breaker, it will send a no-trip signal to the upstream circuit breakers so that the downstream breaker, which is nearer to the fault, should trip immediately and proper time discrimination would be achieved. The advantage of zone protection system is that precise selectivity is obtained and stress during the fault is less.

Refer to Fig. 3.9. Suppose fault takes place in O zone, circuit breaker D will trip in shortest time say, 80 millisecond while circuit breakers B and A will trip after 150 millisecond and 300 millisecond. If fault takes place in Zone P, the circuit breaker B will trip in 80 millisecond instead of 150 millisecond and A will trip in 150 millisecond. Similarly, if fault takes place in zone Q, circuit breaker A will trip in 80 millisecond. In this way, an intelligent discrimination is achieved between the circuit breakers.

Measuring devices on the protective and switching device can also be provided so that a circuit breaker can also display the current and voltage of the system.

MCCB having fast disconnection of fault and current limiting feature will be more and more used in future. Nowadays MCCB with high current and short circuit rating are available in the market.

Incomer (HT)

Earlier, bulk oil circuit breaker and minimum oil circuit breakers were popular because the vacuum and SF6 technology was not fully developed. After the development of vacuum and SF6 technology, these circuit breakers are preferred over the oil type. The reason is that they are maintenance free, occupies less space, reliable and safe. Presently, as per rough estimate, 85% vacuum circuit breakers are used up to 36 KV voltage. Oil-circuit breakers are no more manufactured. SF6 type CB is more popular in higher voltage range i.e. above 66 KV.

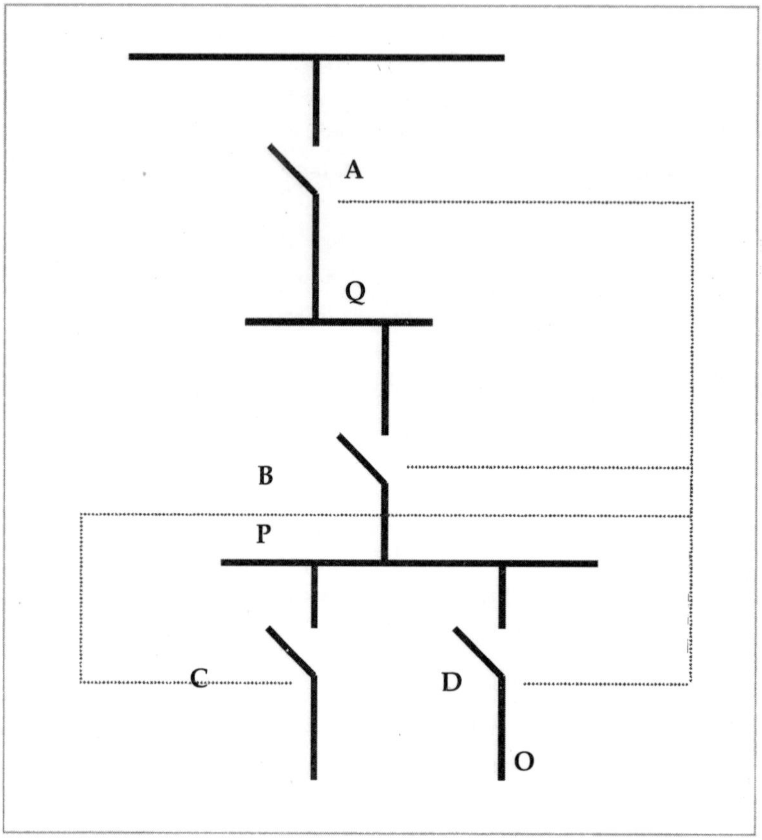

Figure 3.9: Zone protection system

Outgoing Feeders

Motor feeders

As mentioned earlier the trend is FUSEFREE PCC and MCC. The problem that is anticipated in using MCCB in motor feeder is mainly coordination. The apprehension that one may end up using large contactor and bimetallic overload relay. The panel builders who use components of different make have difficulty in establishing the proper coordination. As per IS 13947 (part 4) 1993 type 1 and type 2, two type of coordination are specified. In type 1 coordination, when the fault takes place in motor starter circuit, the components used in the motor starter circuit like contactors, fuses, thermal relays etc. shall not cause any danger to installation or person, but they may not be suitable for further use without replacing or repairing the components i.e. the complete damage to overload relay, contactors etc. are acceptable. Normally, type 2 coordination is more popular since one does not have to replace the components. With current limiting MCCB type 2 coordination can be effectively implemented.

Control

The use of Programmable Logic Controller (PLC) in electrical control system is gaining momentum. A substation may consist many PCC/MCC, transformers, HT switch gears and DG sets for emergency power requirements. Till now different relays for different equipments were being used. For example, for auto starting of DG sets on failure of power supply and auto stopping on restoration of power supply, a numbers of relays, timers and contactors are required. Similarly, ON-load tap changer in a transformer needs relays timers and contactors, Control Cabling and wiring. By installation of PLC number of relays λ, wiring and cables can be reduced to a large extend. Further, the flexibility of system increases. There are three ways by which PLC system can be implemented:

- By providing a single PLC for all the devices with remote terminals for other devices, which are located at a distance from the location of the PLC.
- Individual PLC for controlling a group of switchgears, transformers etc.
- Separate micro-controller in each feeder.

While designing the system, different aspects including the cost of the system are taken into account. One type of scheme suitable for one system may not be suitable for another. If we take cost factor into account, the single PLC will be cheaper, however if the system is large enough, then more than one PLC may be required. Sometimes hot standby PLC is also required.

General Design

It is mentioned earlier that the trend is compactness of the PCC. The circuit breaking devices, contactors etc. are becoming smaller in size. Due to reduction in the sizes of these devices the size of the PCC and MCC also gets reduced. The present practice is to provide modular type, draw-out type outgoing feeder. The feeders of equal rating have equal sizes for interchangeability. The feeders are provided in horizontal one above the another in a multi-tier formation. Plug-in type MCCB will help in making the design more compact. Further, more flexibility can be obtained if the construction of PCC and MCC is changed from welded to bolted type. Any last minute change in the bolted type of construction can be easily done while it becomes difficult in case of the welded type of construction. It would be much better if some standardization is done depending on the rating of the modules of outgoing and incoming feeders. Similarly, standardization in the bus-duct should be done.

Protection Relays

Use of electromechanical relays was very common in the past. Even today we find the mention of such type of relays in specifications of switchgear system. These relays are perfectly right and robust in their functions. However, with the advent of the microprocessor, the entire scenario has changed. The modern microprocessor-based relays are much more superior in many aspects than the traditional electromechanical relays. Some of the advantages are mentioned below:

- Many different types of protective relays have been integrated in one compact unit. In electromechanical relays separate relays are used for separate functions, which occupy a lot of space.
- A very wide range of settings and facility of curve setting for a particular application.
- Highly accurate, measures true r.m.s value.
- Continuous self supervision.
- Remote control due to communication facility with the computer.
- Small power consumption.
- Fault data recording and analysis of fault is possible.
- Maintenance free.
- Display of various electrical parameters.
- Immune to harmonics present in the system.

Another important feature which may be implemented in future PCC/MCC is the detection of arc inside the panel. The arcing takes place due to direct short circuiting on the bus-bar. Microprocessor-based device has already been developed, which detects the arc light and also senses the over-current. On fulfilment of both the conditions, it trips the circuit breaker immediately.

Electrical Parameter Measuring Devices

Mechanical type energy meters for measurement of koliwatt-hour and other energy parameters were being used in the past. The present trend is to use microprocessor-based power monitor, which measures a lot of parameters and stores it. These devices have the facility to communicate with the computer. With the help of special software, reports of the power consumption on hourly basis, daily, weekly and monthly basis can be generated. Future trends can be displayed in a graphical manner.

3.8 Conclusion

In this chapter, the power control centre, bus-bar, selection criteria for HT and LT devices, specifications etc. have been discussed. A comparison between different circuit breakers are given to have a fair idea of HT and LT switch gear. In the example, a step by step procedure for fault calculation, selection of HT and LT equipments, development of the specifications, relay coordination have been given.

3.9 Software

Objective

- Calculation of fault current and development of specifications of HT and LT system.

How to use it

- Click on the "Fault Calculation" from the main menu to open the fault current calculation screen (Refer to Fig. 3.10).

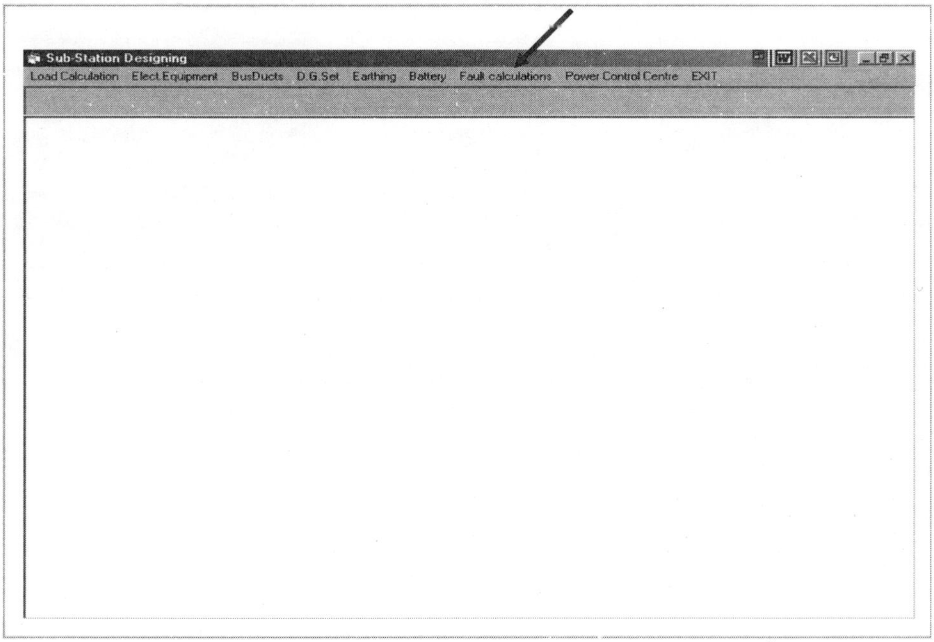

Figure 3.10: Main menu screen

- Fill in the required parameters in the relevant boxes (Fig 3.11).
- Different fault current i.e. L-L, Three phase, L-E, LL-E etc. at three different locations can be calculated. The calculation can be done for the system as shown in the diagram on the screen.

Example 3.3

In Example 3.1 manual calculations are done for symmetrical fault current at three different locations. Fig 3.11, shows the computer screen for the same calculations.

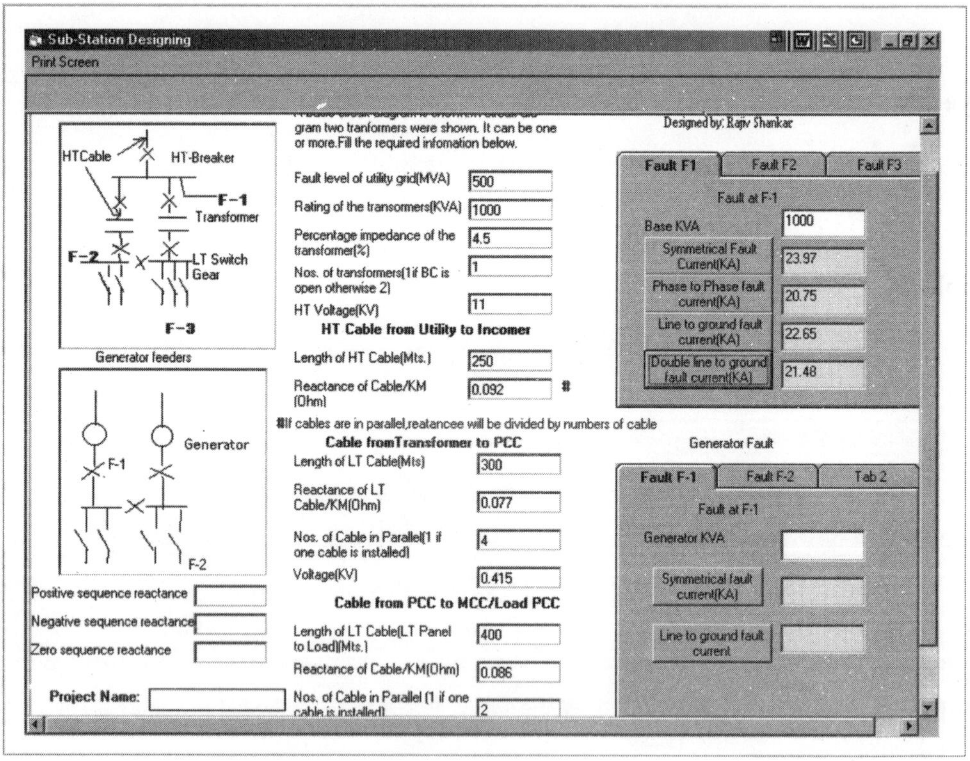

Figure 3.11: Fault current calculation screen

Fault at F2

Per unit reactance X1 = 0.0807 (Positive sequence) as per Example 3.2

Per unit reactance X2 = 0.0807 (Negative sequence, since the negative and positive sequence reactance of cables and transformers are equal. The zero sequence reactance of cable is 3 to 5 times of the positive sequence reactance. We take it as three times.

$X0 = (0.002 + 3x0.00019 + 0.045 + 0.0335\ x3) = 0.148$ per unit

Now all the reactance are converted in ohm unit by using the following formula.

Ohm = per unit reactance x 1000 x KV ^{2}/ base KVA

X1 (ohm) = X2 (ohm) = 0.0807 x 1000 x 0.415^2 /1000 = 0.0138 ohm

X0 (ohm) = 0.148 x 1000 x 0.415^2/1000

= 0.0255 ohm

Line to ground fault

= 3 x E phase- N/(X1 + X2 +X0) = (3 x 240)/(0.0138 + 0.0138 + 0.0255)

= 13559.32 Amp = 13.56 KA

Line to line fault

= 1.732 x E phase-N/(X1 + X2) = 1.732 x 240/(0.0138 + 0.0138)

= 415 /0.0276 = 15036.23 Amp = 15.036 KA

Double line to ground

= X2 x 3 x Ephase –N/(X1 x X2 + X0(X1 + X2)) = 0.0138 x 3 x 240/(0.0138^2 +

0.0255 (0.0138 + 0.0138)

= 9.936/0.00089 = 11164.04 amp = 11.16 KA

Three phase fault

= E phase-N/X1 = 240/0.0138 = 17.39 KA (We have already calculated 17.24 KA in Example 3.1 by per unit method)

Refer to Fig. 3.12, which shows the computer screen of the software for the above calculations.

Fault at F3

Per unit reactance X1 = 0.0807 + 0.0998 = 0.1805 (positive sequence) as per Example 3.1, step 8 Per unit reactance X2 = 0.1805 (negative sequence, since the negative and positive sequence reactance of cables and transformers are equal. The zero sequence reactance of cable is 3 to 5 times of the positive sequence reactance. We take it as three times.

X0 = (0.148 + 0.0998 x3) = 0.447 ohm

Now all the reactance are converted in their ohm values by using the following formula.

Ohm = per unit reactance x 1000 x KV ^{2}/ base KVA

X1(ohm) = X2 (ohm) = 0.1805 x 1000 x 0.415^2 /1000 = 0.03108 ohm

X0 (ohm) = 0.447 x 1000 x 0.415^2/1000

= 0.077 ohm

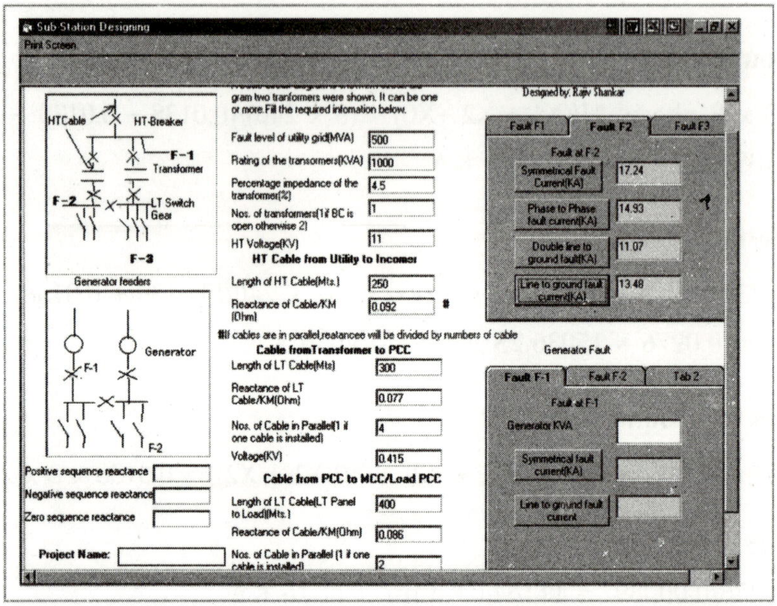

Figure 3.12: Screen for fault current calculation at fault F2

Line to ground fault

= 3 x E phase- N/(X1 + X2 + X0) = (3 x 240)/(0.03108 + 0.03108 + 0.077)

= 5173.9 Amp = 5.17 KA

Line to line fault

= 1.732 x E phase-N/(X1 + X2) = 1.732 x 240/(0.03108 + 0.03108)

= 415 /0.06216 = 6676.32 Amp = 6.676 KA

Double line to ground

= X2 x 3 x E phase –N/(X1xX2 +X0(X1+X2)) = 0.03108 x 3 x240/(0.03108^2 + 0.077(0.03108+0.03108)

= 22.37/0.00575 = 3888.89 Amp = 3.89 KA

Three phase fault

= E phase-N/X1 = 240/0.03108 = 7.722 KA (We have already calculated 7.75 KA in the example by per unit method)

Refer to Fig. 3.13 which shows the computer screen of the software for the above calculations.

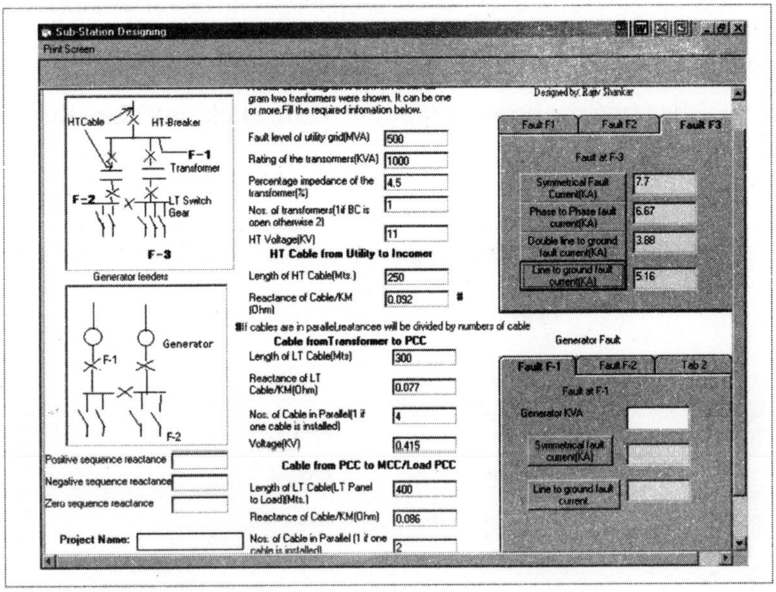

Figure 3.13: Screen for fault current calculation at F3

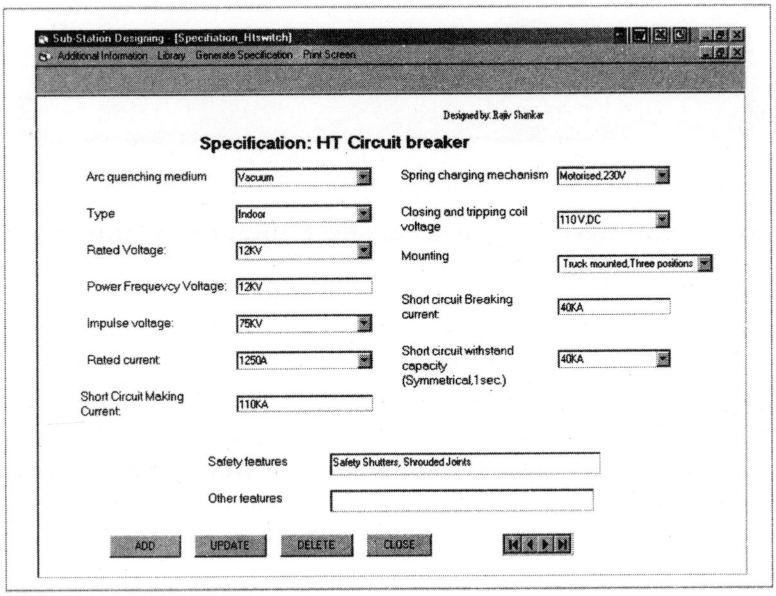

Figure 3.14: Screen for generation of HT circuit breaker specification

- After calculation of fault current for HT and LT system the specifications are developed.
- To generate specifications, click on the electrical equipment from the main menu.
- Select the equipment whose specifications are required to be generated. Sample Specifications of HT circuit breaker is generated by the software as Fig. 3.14 and Fig. 3.15.
- Fill in or select the required parameter against each title on the screen. Before filling the details "ADD" button must be clicked.
- To generate the specifications in tabular form, click on "Generate Specification" on the menu bar.

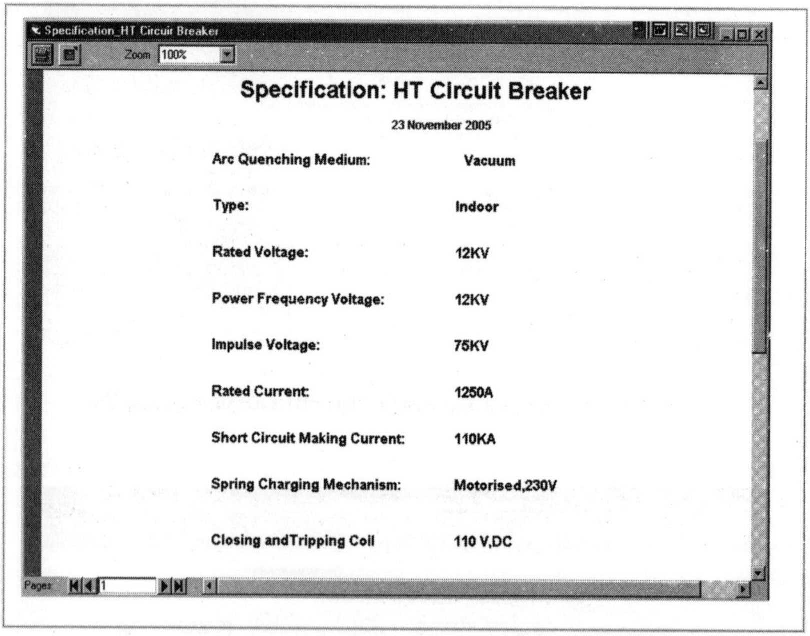

Figure 3.15: Screen for generation of specification in tabular form

Library

The software also consists of a library to store data of circuit breaker. User can store, retrieve and delete the data as per his requirement. The data can be stored by clicking on the "Library Entry" on the menu bar and entering the respective parameters in the specified boxes after clicking on "ADD" button The information stored in the library can be retrieved by clicking on "Library" on the menu bar. Refer to Fig. 3.16, which shows the computer screen of library. Fig. 3.17 shows the screen for entering information in the library.

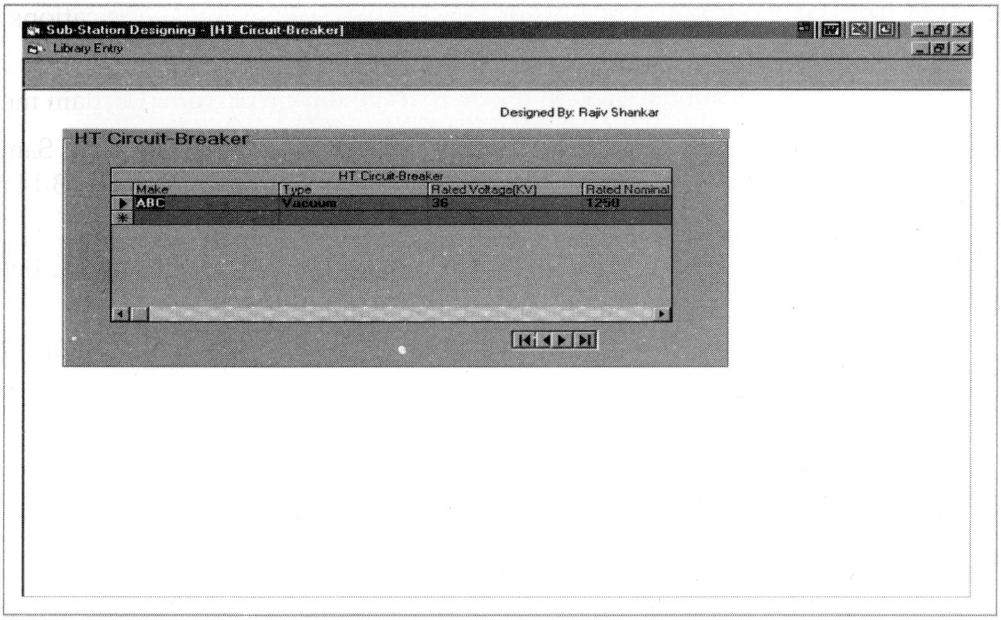

Figure 3.16: Library screen

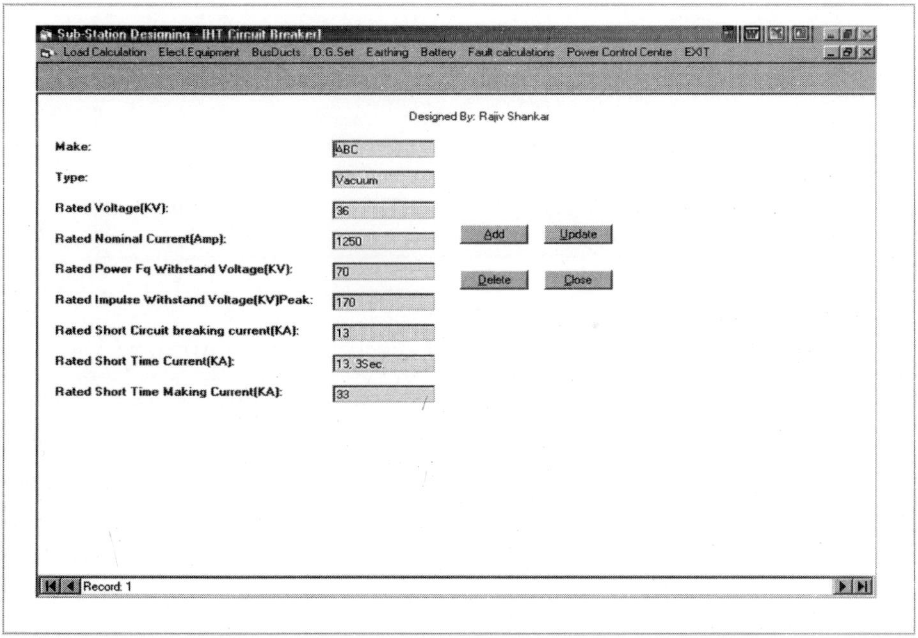

Figure 3.17: Screen for entering information in the library.

4 HT and LT Cables

4.0 Introduction

In any electrical system cables play a very important role in transmission of electrical power from one end to another. It can carry a small load of a lamp and also large load of a generator transformer. Of course, the size of the cable for both types of load will be different. It is also capable of carrying power at high voltage. The cable must be selected and installed in such a way that it should be able to carry the load at the required voltage without damage to itself or the other connected equipments. This aspect will be discussed in more detail in this chapter. The cables must be installed in such a manner that it poses no danger to the person who may come in contact with the cable. In high voltage cable care should be taken that the induction level should not be so high that it develops the hazardous conditions to the nearby equipments. Further care should be taken about the environmental conditions where the cable is going to be installed. For example, the temperature rating of the cable must suit the ambient condition.

In this chapter we would highlight the following:

- Construction of LT and HT PVC and XLPE cables.
- Specification.
- A comparison between PVC and XLPE cable.
- Selection.
- Installation.
- Economics.
- Tests.

4.1 Construction

The most commonly used materials for cable is copper and aluminum. In power cable aluminum is generally used, while in control cables copper is used. Copper has better conducting properties than aluminum. To carry the same current, the aluminum conductor will take 1.6 times more cross section area than the copper conductor. Therefore, the size of aluminum conductor cable will be more than the copper conductor cable. This will result in more space requirement for the aluminum cable. The aluminum cable needs larger surface area for termination. The termination of aluminum cable with the equipment having copper terminal causes bi-metallic effect and formation of aluminum oxide on the metal surface. Using bimetallic washers/ sheet in-between solves this problem. Since copper is not available in sufficient quantity, aluminum cable is used in power circuits.

The power cable consists of aluminum of electrical grade purity and shaped conductor. It may be solid or stranded. In high tension cable the conductors are stranded, compacted and round. Shaped conductors are not preferred in high tension cable since the electrical stresses are higher at the edges of the HT cables. The control cable consists of annealed high conductivity copper. All conductor conforms to IS:8130. Normally, the aluminum conductor upto size 10 sq. mm are solid circular in cross section and sizes above 10 sq. mm are stranded. Single and twin core cables upto 50 sq mm are circular in shape, while in three and four core cables, conductors of cross section area of 25 square mm and above are shaped.

Insulation

A wide variety of insulation is provided for the cable, which depends on the application, installation and environmental conditions. For example, rubber insulation is provided for flexible moving cable. Other insulating material are PVC (Poly Vinyl Chloride), XLPE(Crossed linked poly ethylene), paper etc.

For fixed installation power cables, PVC and XLPE are the most common type of insulation. In HT cable the XLPE insulation is provided in its natural colour since adding colour for core identification deteriorates the electrical properties. The insulation should be free from air voids and foreign material as presence of these material results in unequal distribution of electrical stress. The properties of XLPE is better than the PVC as shown in the Table 4.1. The insulation and outer sheathing compound conforms to IS 5831.

Table 4.1

Comparison between XLPE and PVC cable

Sl. No.	Characteristic	PVC	XLPE
1.	Conductor temperature rating	70^0 C	90^0 C
2.	Life		Longer than PVC
3.	Maximum temperature limit under short-circuit	160^0 C	250^0 C
4.	Overloading	10%	60%
5.	Moisture resistance		100 times more than PVC
6.	Insulation resistance	High	1000 times that of PVC
7.	Weight		Less than PVC

Presently, XLPE insulated cables are preferred over the PVC in HT as well as LT installation due to the above-mentioned qualities of the XLPE insulating material. The dielectric loss of XLPE insulation is very low and does not vary much over the entire operating range. XLPE insulated cable can carry much higher current for the same cross section area in comparison to the PVC insulated cable. Cable with XLPE insulation has also higher short circuit withstand capacity than the PVC cable since it can withstand higher temperature rise in comparison to PVC insulated cable. The XLPE cable can be overloaded upto 60% in case of emergency due to failure of one cable where cables are installed in parallel, so the current carrying capacity will not be hampered.

Inner sheath

In LT cables a common covering is applied over the core of extruded PVC compound or by wrapping of plastic to provide bedding for the armour. In HT cables both conductor and insulation shall be provided with extruded semi-conductor compound shielding. The XLPE insulation is weak to tangential electrical stress and hence for better performance the electrical stress should be radial. The main purpose of providing the semiconductor compound shield is to make the electrical stress radial. In addition to that, in HT cable the insulation is provided with semiconductor tape and non-magnetic metallic tape screen over the extruded insulation to reduce the electrical stress and to carry both leakage and earth fault current. Common inner sheath of extruded PVC is provided. Fillers of non-hygroscopic material is provided.

Armouring

The armouring consists of flat or round galvanized steel wire. For single core cable non-magnetic hard drawn round or flat aluminum wire is provided.

Outer sheath

The tough outer sheath of PVC compound is extruded over the armour/non-magnetic metallic part of insulation/inner sheath (un-armoured cable). In FRLS cables (flame retardant low smoke) special material is used in the outer sheath.

The following tests are performed on the FRLS cables.

- Smoke density measurement
- Oxygen index test
- HCL liberation test

4.2 Specifications

While ordering cable, the parameters as shown in Table 4.2 should be specified:

Table 4.2

Sl. No.	Technical Parameters
1.	System Voltage (KV)
2.	Core size and numbers of core
3.	Grounded/Not grounded system
4.	Conductor material
5.	Type of armour
6.	Type of insulation

4.3 Selection

The important parameters as shown in Table 4.3 should be taken into account while selecting the cable.

Table 4.3

Sl. No.	Parameters
1.	Voltage
2.	Type of system
3.	Maximum load current
4.	Short circuit current and duration

5.	Future load requirement
6.	Protection against mechanical damage, excessive exposure inflammable environment
7.	Installation conditions (air, duct, ground)
8.	Proximity with another cable
9.	Maximum ambient temperature
10.	Depth of laying
11.	Statutory regulations of Electricity Boards or any other agency
12.	Voltage drop
13.	Economic consideration – losses in cable

4.4 Tests

The following routine tests are carried out

- Conductor resistance test
- High voltage test
- Partial discharge test (in case of HT cables)

The test voltages as shown in Table 4.4 are applied between conductor and screen/ armour.

Table 4.4

Voltage rating	Test voltage
1.1 KV (E)	3 KV
3.8/6.6 KV (E)	12 KV
6.35/11 KV(E)	17 KV
11/11 (KV)(UE)	28 KV
12.7/22 KV(E)	32 KV
19/33 KV(E)	48 KV

The above voltages are applied for five minutes.

4.5 Installation

The installation of cable shall be done as per IS: 1255. If installed in air, cable trays of adequate size shall be provided. Cables shall be properly clamped on cable trays. If the cables are installed in group and the distance between the two cables is 75 mm (in a horizontal plane) no de-rating is required upto 70 sq mm cable and above 70 sq mm the distance should be 150 mm. If it is less, then de-rating is required. The bending radius should be 15 times the diameter of the cable. In case of 1.1 KV multi-core cable, it can be 12 times the diameter of the cable. Further, as mentioned above, cables are selected according to the environmental conditions. The current ratings are based on the normal/standard conditions of installation as mentioned in Table 4.5

Table 4.5

	PVC Cable	**XLPE Cable**
Maximum conductor temperature	70⁰C	90⁰C
Ambient air temperature	40⁰C	40⁰C
Ground temperature	30⁰C	30⁰C
Depth of laying	75 cm	90 cm
Thermal resistivity of soil	150⁰C cm/watt	150⁰C cm/watt

If the cables are installed in such a way that they are exposed to the sun, then the rating of the cables will get reduced in comparison to the cables which are shielded from the sun.

Short circuit ratings of the cables are specified by the manufacturer for a temperature rise of 250⁰C (XLPE) and 160⁰C (PVC) for one second. If the cables are subjected to more than one second the short circuit rating as specified for one second should be divided by $\sqrt{t}$. Where *t* is the time for any other short circuit duration.

Proper care should be taken while terminating the cable.

- Correct size of lug and also of correct material shall be used. Crimping of lugs must be done by the proper crimping tool.
- Sufficient clearance should be maintained in the terminal box where cable is to be terminated.
- Nut-bolts and spring washers of adequate size should be used for properly tightening the lugs to the terminal strip. Looseness of contact will result in heating and burning of cable insulation and terminal strip, and sometimes it results in short circuiting.
- Segregation of cable should be maintained, i.e. power cable, control cable and communication cable should be separated from each other.
- If cables are bunched together, then it should be de-rated by applying appropriate de-rating factor.
- Cable tags and core identification numbers are provided, which will be helpful in maintenance of the electrical system.
- If cables are laid underground then sand for cushioning effect and bricks for mechanical protection should be provided. Hume pipes of proper size shall be provided where cables are supposed to cross the road, drain etc.
- Cable route marker at appropriate distances shall be provided for underground cables.

4.6 Economics in Selection of Cable

It is well known that due to resistance of cable, I^2R losses take place in the cable when current flows through it. If the size of the cable increases, the resistance reduces and therefore the losses also get reduced. Due to reduction of losses, the consumption of electrical energy as well as the electricity charges reduces. However, as the cable size increases its cost also increases. The cable is selected so that the combined effect is minimum. It is required that while selecting cable for a particular application calculation should be done for selecting cable of most economical size. The cable which has minimum pay-back period should be selected. This concept will become clear from the worked examples given in this chapter. Further, to do a quick calculation, the software attached with the book should be used.

IS standards

XLPE cables: IS 7098

PVC Insulated heavy duty power cable IS:1554

PVC Insulated control cable IS:1554

PVC Insulated cable IS:694

Voltage drop during starting of a motor

Normally, when an induction motor is started with DOL (direct on line) starter, it takes starting current equal to 6 to 7 times of its rated current. This large starting current comes down to its normal value as the motor picks up the speed and reaches the rated speed. The large starting current causes drop in voltage at motor terminal. Therefore, the cable should be selected in such a way that the drop remains within the limit which is 15% of the system voltage. If more than 25% voltage drop takes place during starting of motor, there are chances that the contactor used in motor starting circuit may drop (Refer to Example 4.3).

Example 4.1

Refer to Table 1.14 of Chapter 1, which is load table for steam generation plant of a process industrial unit. We take an example of providing LT cable/cables for air-conditioning plant. From the load table we find the following air-conditioning load.

- Chiller compressor: 88 KW
- Air handling, cooling tower and miscellaneous other load: 40.8 KW

  Total Load: 128.8 KW

Other details are as follows:

- Type of cable installation: Underground

- Depth of laying underground cable: 750 mm
- Ambient temperature: 40^0C and ground temperature 35^0C
- Length of underground cable: 250 metres.

With the above data the size of LT cable is required to be calculated.

The first step is to select the size of cable for 128.8KW or say 129 KW load. This load is three phase load and the current will be

- Current = 129/(0.85 x 0.415 x 1.732)

  = 211.14 amp.

For the above mentioned current, let us take a three and a half core cable of cross-section area of 240 square mm. Now the question is whether this size is adequate or cable of larger size is required, and if larger size cable is required, then again the question is what should be the size and what will the economics be of choosing a larger size cable? The answer depends on the following factors/parameters of cable.

- De-rating factors
- Short circuit capacity
- Voltage drop during starting of chiller compressor motor

The current carrying capacity of 240 square mm cable will be 280 amp. Cable is to be laid underground. No de-rating is required if the ground temperature is 30^0C; we take ground temperature as 35^0C. At this temperature a de-rating factor of 0.94 is applied. Depth of laying is 750mm and only single cable is to be installed, therefore the de-rating will be 280 x 0.94 = 263.2 amp.

Voltage drop calculation

Resistance at 70^0 C = Resistance at 20^0 C x (1 + 0.004031 x (70-20))

= 0.125 x (1 + 0.201)

= 0.1502 ohm/Km

Resistance of 250 mtrs cable = 0.1502 x 250/1000 = 0.03755 ohm

Reactance of 250 mtrs cable = 0.087 x 250/1000 = 0.02175 ohm

The voltage drop will be = 1.732 x Current (cable resistance x cosϕ + cable reactance x sinϕ)

= 1.732 x 211.14 (0.03755 x 0.85 + 0.02175 x 0.526)

= 15.85 V

4.7 Conclusion

From the above discussion it is clear that the following points are very important while selection of cable is made:

- Calculation of de-rating factor
- Short circuit capacity of cable
- Voltage drop in cable
- Economics in selection of particular size of cable

Short circuit capacity of cable is particularly important in case of HT cable. It also depends on the time taken in clearing the fault i.e. if any fault takes place in the HT system, it should be cleared in shortest time so that the cable is not damaged.

It is also important to select cable of such size that the voltage drop should be within the limit. If voltage drop during the starting of motor is more than the specified limit, then the motor may not get the required torque for starting.

Sometimes, it is cheaper to take a cable of higher size, though the capital cost of that cable may be high; but if the total cost including the cost of losses is taken into account, the cable of larger size is found to be cheaper than the cable one size smaller.

4.8 Software

Objective

- Selection of HT and LT cable based on load and short circuit rating.
- Calculation of de-rating factor.
- Calculation of voltage drop in the cable.
- Calculation of voltage drop during starting of a motor.
- Calculations for selection of the most economical cable size by taking into account the losses.
- Development of specifications for HT and LT cables.
- Storage and retrienval of cable parameters from the library.

How to use it

- Click on the "Electrical Equipment" on the main menu.
- Select LT or HT cable as the case may be. (Refer to Fig. 4.1)
- The de-rating of cables, voltage drop as calculated in Example 4.1 can be calculated by the software.

- Select single core or multi core from the sub menu.
- Select the method of installation of cable i.e. underground, on duct or on cable tray from the submenu.
- Fill in the required data, and the de-rating factor can be calculated. (Fig. 4.2 & Fig. 4.3)
- Similarly the voltage drop in the cable can be calculated. (Fig. 4.4).

Economics in selection of cable can be calculated after clicking on the "Economics in

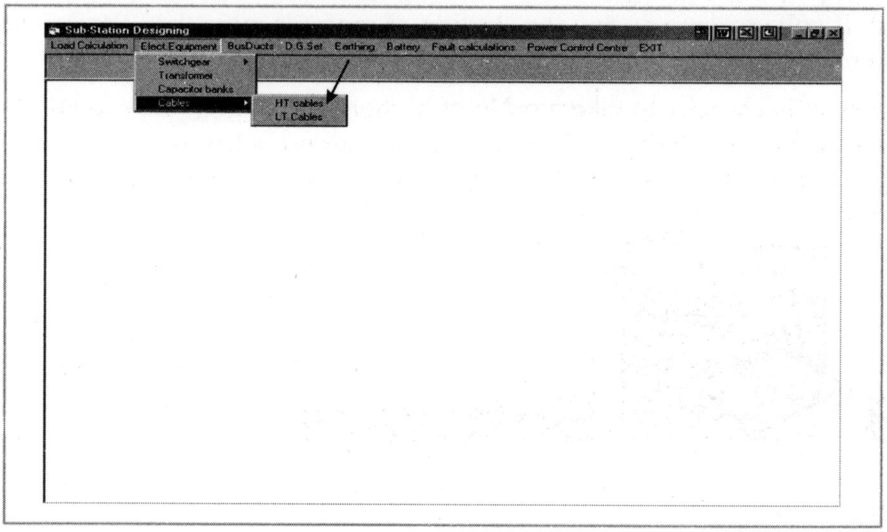

Figure 4.1: Screen showing main menu bar and sub menu.

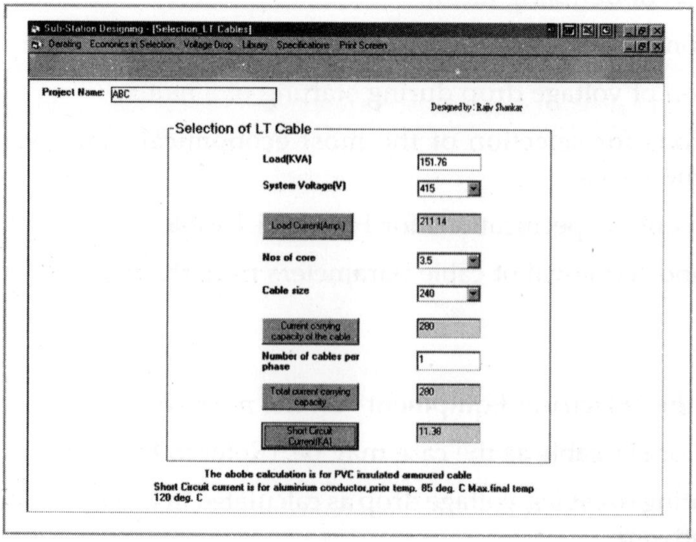

Figure 4.2: Screen for selection of cable.

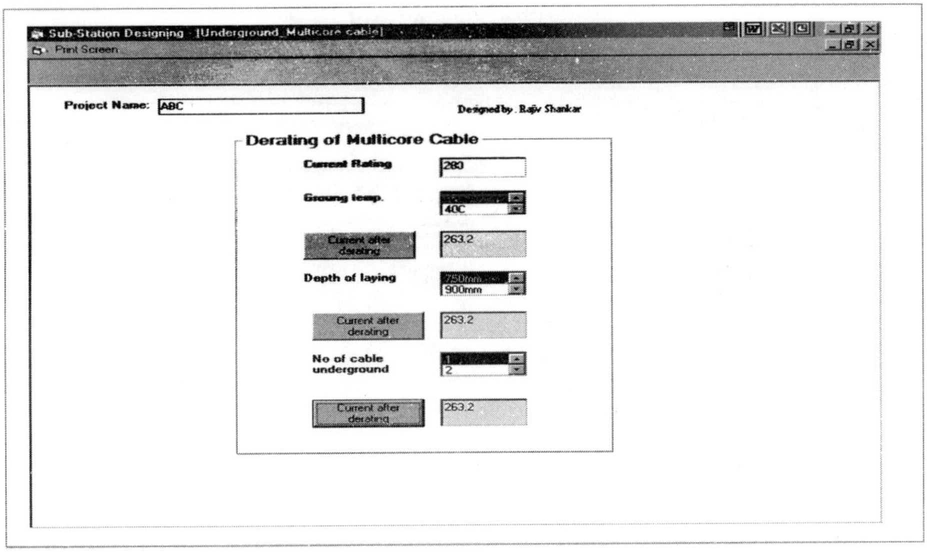

Figure 4.3: Screen for calculation of derating of cables

Selection" on the menu bar. The respective screen opens up (Fig. 4.5). The data in the specified boxes are filled up, and the result is obtained after clicking the respective buttons. Refer to Example 4.2 for calculations.

Example 4.2

In Example 4.1, we have selected a cable of 3.5 core and 240 sq mm and of aluminium conductor for 211.14 amp, and also calculated the de-rating and voltage drop. After de-rating, the current carrying capacity is 263.2 amp and voltage drop is 15.82 V.

Now to find out the economics in selection of cable, we have to calculate the total cost, i.e. cost of losses and the cost of cable of different sizes of cables.

In the following example, first we will calculate the total cost of 3.5 core 240 sq mm cable and then compare it with the 3.5 core, 300 sq mm cable.

Total copper losses in the cable will be

Demand Factor ^{2} x (I^2 R) x (Loading Hrs x unit charges) + (Demand Charges)

I = Current through cable

R = Resistance

Demand factor = 0.67

I = 211.2

R = 0.0375

Unit charges = Rs. 4.5

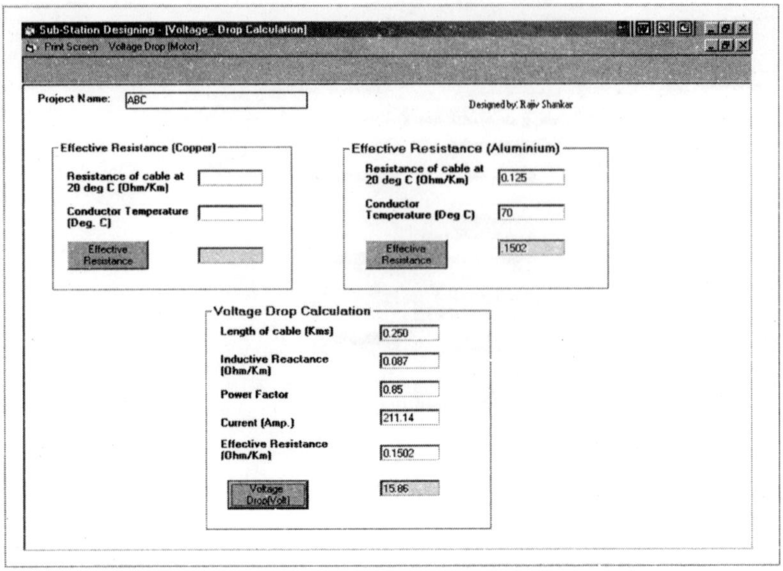

Figure 4.4: Screen for calculation of voltage drop

Demand charges = Rs. 250/ month/KW, Rs. 3000/annum/KW

Cost of cable = Rs. 700/metre

Life of cable = 20 years

Interest = 7%

Depreciation = 3%

Insurance = 0.6%

Cost of Copper losses = $0.67^2 \times \frac{(3 \times 211.2^2 \times 0.0375)}{1000} \times (8760 \times 4.5) + 3000$

= Rs. 91798.69

Cost of 250 mts cable = Rs. 250 x 700 = Rs 175000

Total running cost = Carrying charges x cost of cable/100 + cost of annual copper losses

= 10.6 x 175000/100 + 91798.69

= Rs. 110348.69

Calculation of present value of the above cost

((100/rate of interest) x (1-1/(1+rate of interest/100)^p)) x Annual copper cost

Where p is life of cable in years

= 10.59 x 11.0348.69 = Rs. 1168567.98

Total cost of cable = 1208846.7 + 175000 = Rs. 1343567.98

Now the above calculation is done with 3.5 x 300 sq mm cable

The cost of cable is assumed as Rs. 775/metre and the resistance will be 0.03 ohm for 250 metres cable.

Cost of Copper losses = 0.67^2 x (3 x 211.2^2 x 0.03/1000) x (8760 x 4.5) + 3000

= Rs. 74038.95

Cost of 250 mts cable = Rs. 250 x 775 = Rs. 193750

Total running cost = Carrying charges x cost of cable/100 + cost of annual copper losses

= 10.6 x 193750/100 + 74038.95

= Rs. 94576.45

Calculation of present value of the above cost

((100/rate of interest) x (1-1/(1+rate of interest/100)^P)) x Annual copper cost

= 10.59 x 94576.45 = Rs. 1001564.6

Total cost of cable = 1001564.6 + 193750 = Rs. 1195314.6

Difference in cost 1343567.98 – 1195314.6 = Rs. 148253.37

We find that the cable of higher size is cheaper if the cost of losses for life of cable is also taken into account though the capital cost of 300 sq mm cable is more than the 240sq mm cable. Refer to Fig 4.5, which shows the computer screen for the above calculation.

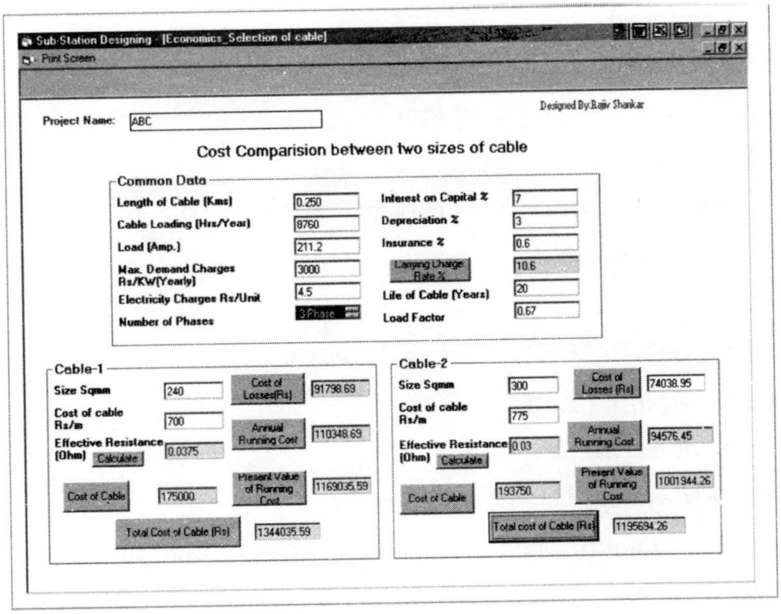

Figure 4.5: Screen for cost comparison between two sizes of cable

- While selecting the cable, the other important calculation is voltage drop during starting of the motor. In industry, the voltage drop calculation is important for large size of motor. Large motors are connected with 3.3 KV or higher system voltage. These motors are controlled with HT circuit breaker and are of direct on-line start type. One dedicated screen is developed for such type of calculation. Click on the Voltage drop (Motor) on the menu bar and the respective screen will open up. Fill in the required data in the respective boxes and click the button "Percentage voltage on the motor terminal" to calculate the percentage voltage at the motor terminal at the time of starting. (Refer to Example 4.3).

Example 4.3

In this example, we take a case of induced draft fan motor. (Fig. 4.6)

Rating: 430 KW

Voltage: 3.3 KV

Rated current: 88 amp

Transformer KVA: 2500

Percentage impedance: 6%

Starting: DOL

Cable length from 3.3 KV switchgear to motor: 350 metres

Cable size: 3 x 300 sq mm

Reactance: 0.092 ohm/KM

Cable length from transformer to 3.3 KV panel: 400 metres

Cable size: 3 nos per phase,1 x 300 sq mm

Reactance: 0.092 ohm/KM

Cable length from 11 KV panel to transformer: 300 metres

Cable size: 1 x 630 sq mm/phase

Reactance = 0.092 ohm/KM

Cable length from grid to 11 KV panel: 600 metres

Cable size: 1 x 630 sq mm/phase

Reactance = 0.092 ohm/KM

Fault level of 11 KV system 500 MVA

Calculation of percentage of rated voltage at motor terminal during starting:

Base KVA = 10000

Percentage impedance of 11 KV grid

= 10000 x 100/500000 = 2%

Percentage reactance of 11 KV cable:

= 0.092 x 600/1000 = 0.0552 ohm

Percentage X = ohm x KVA Base/(10 x KV^2)

= 0.0552 x 10000/(10 x 11^2)

= 0.456 %

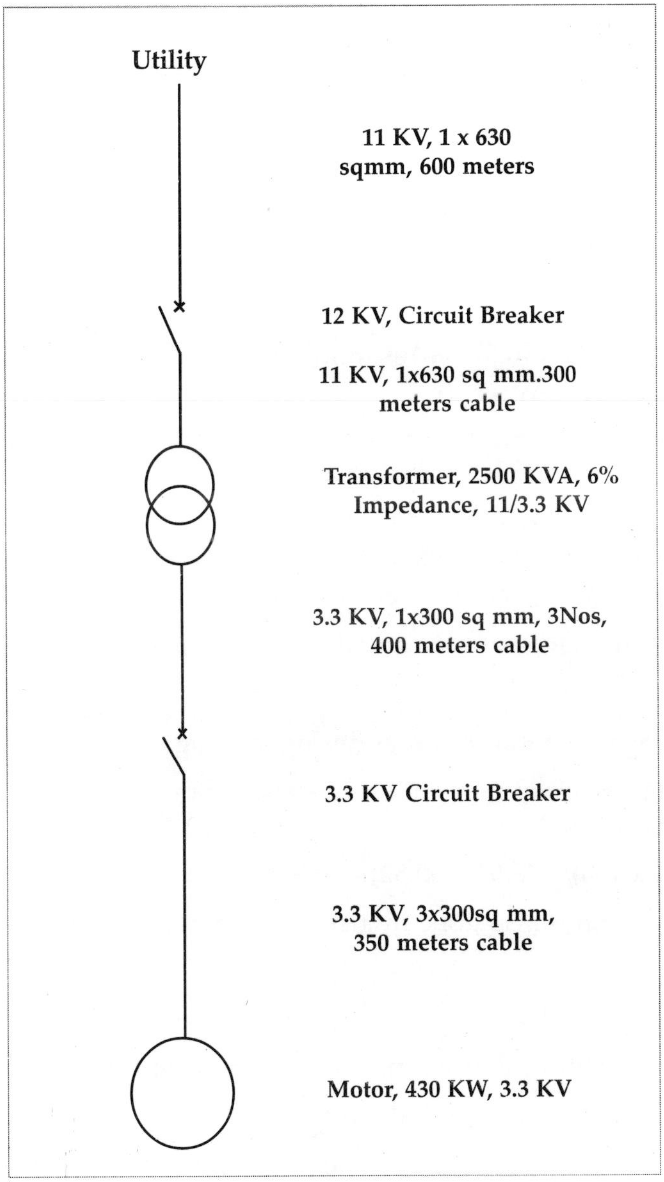

Figure 4.6: Single line diagram of ID fan motor

Transformer to 3.3 KV switchgear:

0.092 x 400/1000 x 3 = 0.0122 ohm

Percentage X = 0.0122 x 10000/(10 x 3.3^2)

= 1.12%

3.3 KV Switchgear to Motor

= 0.092 x 350/1000

= 0.0322 ohm

Percentage X = 0.0322 x 10000/(10 x 3.3^2)

= 2.95%

Motor impedance

Percentage Zm = 100/Il/If

Il = Locked rotor current

If = full load current

Percentage Zm = 100/6 = 16.67% (Assuming Il/If = 6, due to DOL starting)

Percentage Zm on base KVA

= 10000 x 16.6/430/0.9

= 348.91%

Transformer percentage impedance on base

10000 x 6/2500 = 24%

Total percentage reactance of cables and transformer

= 2.0 + 0.456 + 1.12 + 2.95 + 24 = 30.52%

Percentage Voltage at motor terminal during starting

= Percentage impedance of motor/(% impedance of motor + Percentage impedance of transformers and cables)

= 348.91/(348.91 x 100)/(348.91 + 30.52) = 91.9 %

The above calculation is also done by using the software. (Refer to Fig. 4.7)

Specifications

The software is capable of developing specifications of HT and LT cable.

- Click on the specifications on the menu bar.
- Click "ADD" button and fill in the required parameters in the blank boxes. Click print to print the screen. (Refer to Fig. 4.8)

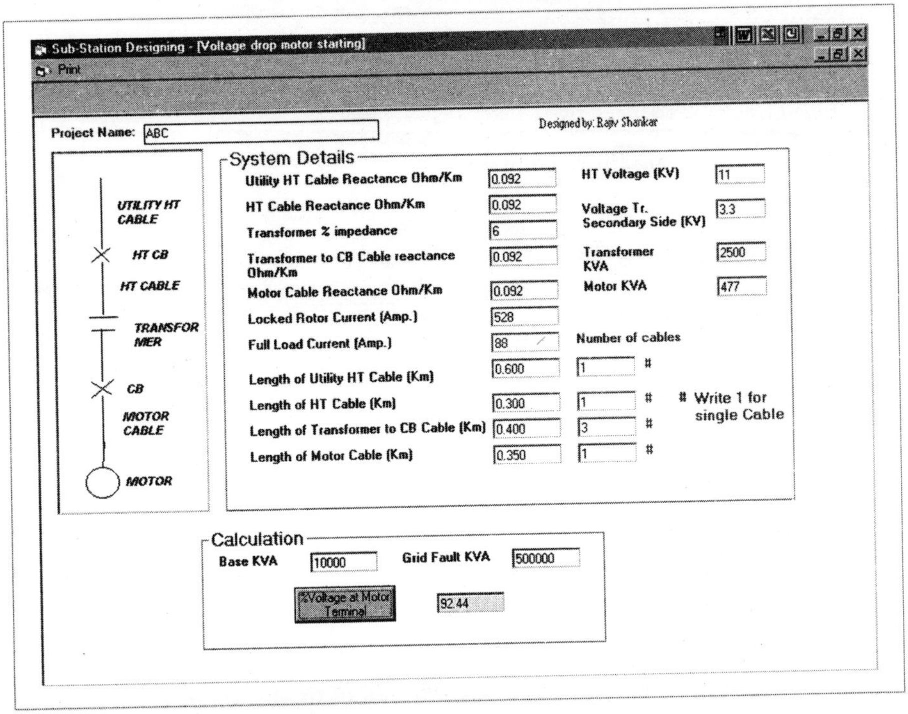

Figure 4.7: Screen for calculation of % voltage at motor terminal during starting of motor

- Click "Generate specification" to develop specifications in a tabular form and the same can be printed also. (Refer to Fig. 4.9)

Library

The data related with cables available in the market of different manufacturers can be stored, retrieved and deleted by the user.

- Click on the Library on the menu bar. The table consisting of the stored data opens up. (Refer to Fig. 4.10)
- To store the data click on the Library entry button, and a screen opens up.
- Click "ADD" button and enter the data to be stored in the library (Refer to Fig. 4.11)

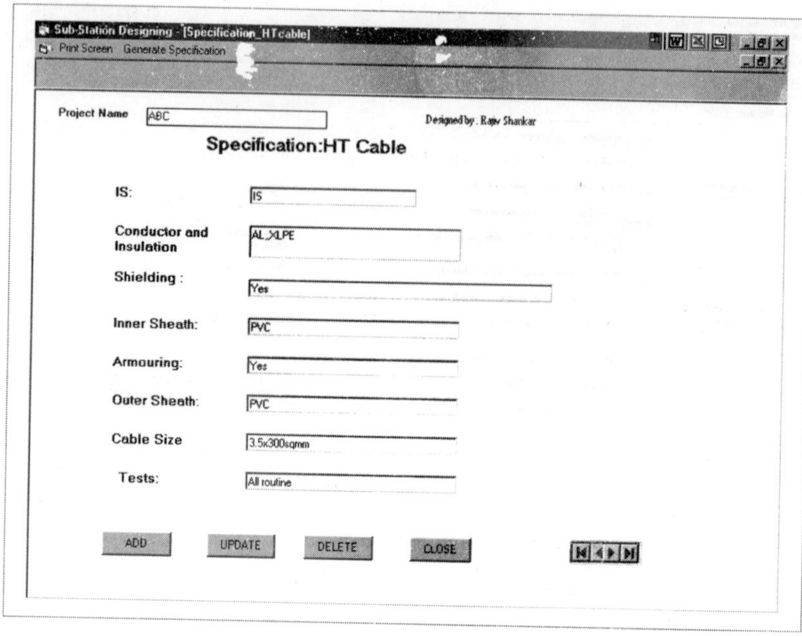

Figure 4.8: Screen for development of specification

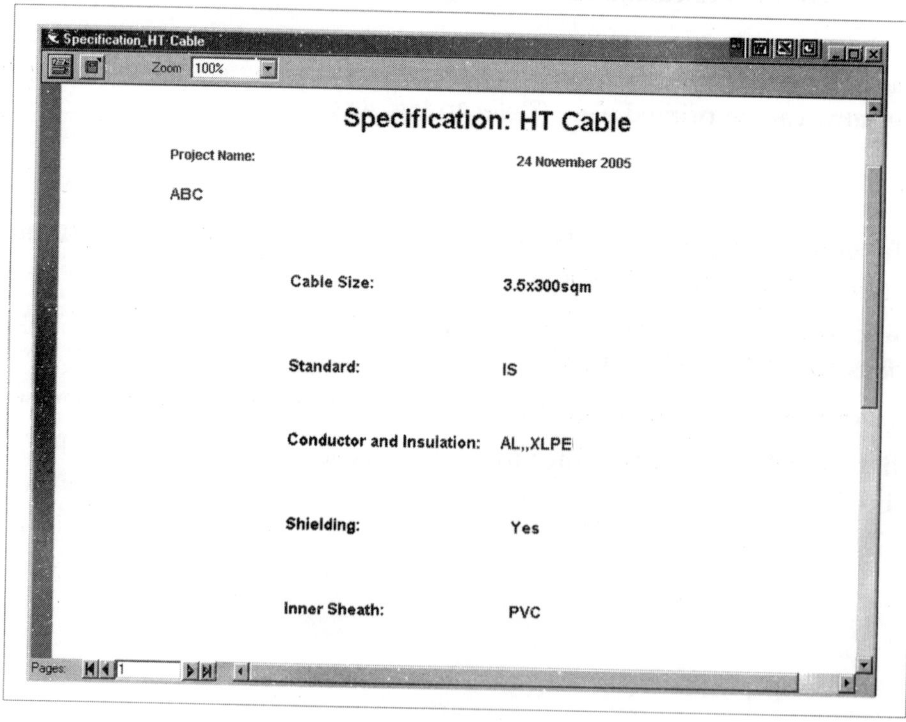

Figure 4.9: Screen for generation of specifications in a tabular form

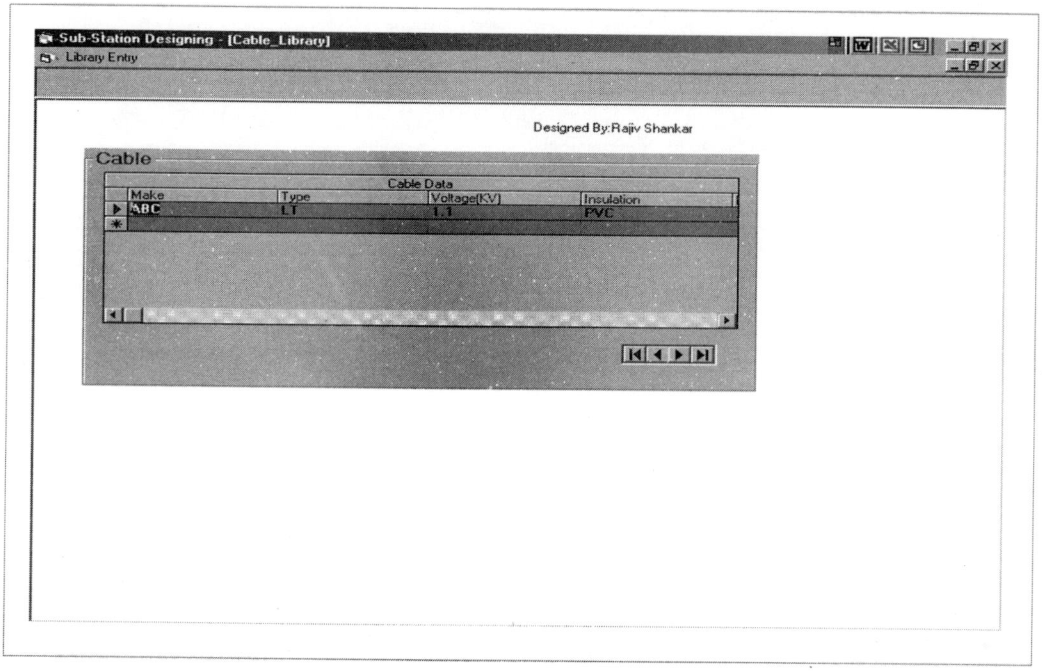

Figure 4.10: Library showing various parameters of cables

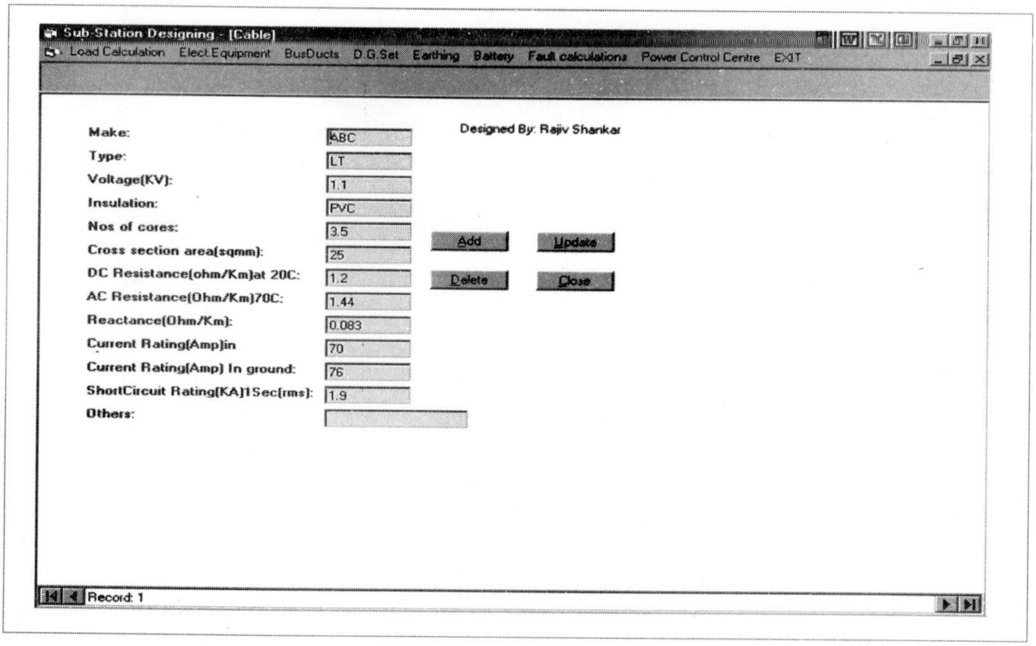

Figure 4.11: Screen for entering information related with cable in the library

5 Bus-Duct

5.0 Introduction

The bus-duct system consists of aluminium or copper bus-bars in a metal enclosure. It is an alternative for cable. Normally, when the large current is to be transferred from one system to another like from transformer to power control centre, from one power control centre to another power control centre, from generator to power control centre etc. a large number of cables are required to be installed in parallel. Installation and termination of large numbers of cables in parallel is cumbersome. It requires more space for installation and termination of cables. To overcome this problem bus-duct is used. The following aspects of bus-duct are covered in this chapter:

- Specifications.
- Sizing calculations.
- Selection of bus-duct.

5.1 Selection

Correct selection of the bus-duct is very necessary for good performance of the electrical system. There are certain points as mentioned below that should be taken care of while selecting of the bus-duct.

- Ambient conditions.
- Material.
- Installation.
- Rating (voltage, current, short circuit rating etc.)

- Enclosure
- De-rating factors
- Termination
- Grounding

In general, bus-ducting seems to be a simple type of equipment consisting of bus-bar in a metallic enclosure. However, it is a very critical equipment in electrical system. Any fault developed in the bus-duct can hamper the total power supply.

5.1.1 Ambient conditions

Among ambient conditions the most important factor is the temperature. As per IEC 694 and IS 12729, the maximum temperature of air insulated aluminum conductor shall be 90°C and not exceed 40°C over the ambient temperature. It may be mentioned that often maximum ambient temperature and reference ambient temperature are considered same. An arbitrary figure is assigned to the ambient temperature. For example, in some of the specifications it has been found that ambient temperature is specified as 45°C or sometimes 50°C too. The electrical equipments are designed on 40°C and if the ambient temperature is specified as 45°C, the vendor has to de-rate the equipment so that it can work satisfactorily at 45°C temperature. It may be clarified that the maximum ambient and the reference ambient temperature are not the same.

As per IS 9676, the standard reference ambient temperature for rating of electrical equipment is 40°C. It is important to give a correct figure related with the ambient temperature in the specifications of the equipment rather than assigning some arbitrary figure of maximum temperature, which may be the maximum temperature for a particular place for a short duration of time.

The other ambient conditions which should be taken care of are the height from the sea level where the bus-bar system is going to be installed. With increase in height, the air becomes thin and the cooling becomes less. The relative humidity is also an important factor.

5.1.2 Material

Basically, two types of material are used for the bus-bars.

- Copper
- Aluminium

Electrical grade E91E aluminium is extensively used as bus-bar material. The reason behind this is that aluminium is easily available and is cheaper in comparison to copper. If we compare the electrical and mechanical properties of copper with those of aluminium, then we will find that copper is much better than aluminium. Copper has got more tensile strength, its conductivity is 61% higher than that of aluminium. Therefore, for the same current rating aluminium conductor requires larger cross section area. Thus the size of the enclosure increases. Where the size of the bus-duct is the main consideration due to the space restriction, copper bus-bar is preferred. Copper is 50% heavier than aluminium.

A hard oxide film is quickly formed on the surface of the aluminium when it is exposed to the atmosphere. This oxide layer is non-conductive. It is necessary to electroplate the surface to make an effective joint. The other important point to be considered is that the various other devices like circuit breakers, switch fuse units etc. are to be connected with the bus-bar and the terminals of these devices are of copper. The joining of two different metals causes bimetallic effect. If moisture is present, then electrolytic action takes place and the joint is corroded fast. The coefficient of thermal expansion of copper and aluminium are different and the joint is likely to become loose after some time. To overcome this problem copper-clad aluminium washers or bus-bars shall be used.

5.1.3 Rating

The bus-duct should be chosen in such a way that it should be able to carry the required current at the system voltage. In addition to this, some spare capacity for future requirement shall also be kept. It should also be able to withstand the fault current during occurrence of fault in the system without any damage for the specified time i.e. one second or three seconds. The current rating of bus-bar depends upon the temperature rise over the ambient temperature. Suppose we say that temperature rise shall be 90^0C over the ambient temperature of 40^0C. This means that the current through the bus bars should be limited so that the temperature should not increase the limit of 90^0C. The heat in the bus-bars is dissipated through convection, radiation and conduction. The heat dissipation through conduction is small, since the bus-bars are supported on the insulating material. Most of the heat is dissipated through the convection and radiation in the surrounding medium. It also depends on the surrounding temperature, mounting of the bus-bars. For example, if the bus-bars are mounted vertically, then the heat dissipation will be more than the horizontally mounted bus-bars. In large industry, it is common to have bus-ducts of more than 2500 amp rating. As the current rating increases, the size of the bus-bar also increases. For large current rating multiple bus-bars are used. The space between them is kept equal to the thickness of the bus bar. In this type of construction the heat dissipation due to radiation as well as convection from the inner portion of the bus-bar becomes less due to screening of the inner bus by the outer bus. Further, the current in the bus-bars is not uniformly distributed (AC application) due to skin and proximity affect. The self and mutual inductance increases due to skin and proximity effect and causes current to be concentrated on the outer surfaces of the conductor. This increases resistance of the outer surface in comparison to the inner surface and further limits the current carrying capacity of the bas-bars. The ratio Rac/Rdc (AC resistance/DC resistance) is the indicator of the distribution of the current through the bus-bars. If the ratio is unity, the current distribution is uniform, but if it is more than unity, and increases, then the non-uniform distribution of current also increases. As the number of bus-bars increases, the proportion of the current carried by the inner bus-bars decreases and a point will reached after which any addition of bus-bars will not cause any appreciable increase in current rating. More than four to five bus-bars in parallel should not be used. Interleaving of bus-bars should be done to solve the problem of the skin and proximity affect. The conductor of one phase is separated and other phases are interleaved i.e. R-Y-B-R-Y-B as against the convention RR-YY-BB. This causes greater uniformity of the inductance and uniform distribution of the current. The phases are adjacent to each other; therefore, more creepage distance is required. Due to this, the size of the bus-duct increases. This is a major disadvantage where size is the limitation.

During the short circuiting, the bus-bars should be able to withstand the thermal as well as mechanical stress. When a short circuiting takes place, the temperature rise is directly proportional to the square of the r.m.s. value of the fault current. The duration of short circuiting is very small i.e. one second till the breaker opens and clears the fault. The heat dissipation through convection and radiation during this short duration is negligible and all the heat is absorbed by the bus-bar itself. The temperature rise due to the fault can be calculated by applying the following formulae.

$T = K(I/A)^2 (1+\alpha\theta)10^{-2}$

T = temperature rise per second

A = Conductor cross-section area

α = Temperature coefficient of resistivity at 20 ^{0}C/^{0}C

= 0.00393 for copper

= 0.00386 for aluminium

θ = Temperature of the conductor at the instance at which the temperature rise is being calculated.

K = Constant, 0.52 for copper and 1.166 for aluminium.

5.1.4 Enclosure

The bus-bars in a bus-duct system should be enclosed in a sheet steel enclosure. The class of protection shall not be less than IP: 45. The enclosure shall be painted after the treatment of the metal with the specified paint. If the length of the bus-duct is large, then the expansion joints should be provided at suitable distances. The bus-duct should be terminated by using the flexible terminal strip of the same current carrying capacity as that of the bus-duct. Normally, the bus-duct is supplied in sections and joined at site where it is to be installed. The bus-bars of different sections are joined by over lapping the bus-bars and bolting or clamping. The contact pressure shall be proper, and the surface area should be cleaned and the overlapped area should be equal to the width of the bus-bar. In bolted joints the contact area of the two joined bus-bars remains almost beneath the head of the bolts since pressure is maximum beneath the bolt head. Fish plates with split washers provide greater contact area and good joint. The maximum heat generation takes place at the joints since the resistance at the joints is more in comparison to the bus-duct length.

Bus-duct should have proper supports at suitable distances. It should have grounding strips at both the sides.

Example 5.1

5.2 Calculation of Bus-Bar Size

Before discussing the procedure for calculation of bus-bar size, it is desirable to have a look at the de-rating or correction factors.

5.2.1 De-rating due to temperature rise

Table 5.1

Temperature Rise	De-rating factor
1. Bus-bar temperature rise 40^{0}C Over-ambient 35^{0}C	0.88
2. Bus-bar temperature rise 30^{0}C Over-ambient 35^{0}C	0.75

5.2.2 De-rating due to enclosure

Table 5.2

Sl. No.	Enclosure	(Cross-section area of bus-bar)/(Total cross-section area of enclosure)	De-rating factor
1.	Outdoor	<1% 5% 10%	0.95 0.90 0.85
2.	Indoor where enclosure is in a well-ventilated room	<1% 5% 10%	0.85 0.75 0.65
3.	Indoor where enclosure is in a poorly ventilated room and the room temperature is high	<1% 5% 10%	0.65 0.60 0.50

5.2.3 De-rating due to altitude

The limit of temperature rise should be reduced by 1% for each 300 metre in excess of 1000 metre above sea level.

5.3 Specifications

The following should be specified while finalizing the specifications:
A specification for a typical bus-duct has been mentioned in Table 5.3

Table 5.3

	Description	Details
1.	Rated Current	1000 amp
2.	Insulating voltage	1000 V
3.	Frequency	50 Hz
4.	Protection degree (Enclosure class)	IP-45
5.	Material of bus-bars	Aluminium
6.	Short circuit rating (KA) peak	60 KA
7.	Short circuit rating (r.m.s.),1 sec.	40 KA
8.	Resistance, milli-ohm/metre	<0.2
9.	Reactance, milli-ohm/metre	<0.15
10.	Impedance, milli-ohm/metre	<0.2
11.	Cross-section area	To be specified by vendor
12.	Grounding strip	25 x 6 mm Al Both sides
13.	Expansion joint	To be provided
14.	Space heaters	To be provided
15.	Painting	Grey
16.	Sleeved conductor	Required
17.	Shrouded joints	Required
18.	Fire barrier	Required
19.	Insulators	Class A
20.	Length of each sections	3 metres approximately

Step-by-step calculation has been shown in Table 5.5. In the remark column that particular step has been explained.
Required data is given in Table 5.4.

Table 5.4

1.	Thermal rating	1150 amp
2.	Fault level	50 KA for 1 sec.
3.	Ambient temperature	40^0C
4.	Permissible temperature of bus-bar after short circuit for one second	200^0C
5.	Bus-bar material	Aluminium alloy E91E
6.	Size of bus-bar	2 x 150 x 6 (Three phase) and 1 x 150 x 6 (Neutral)
7.	Cross-section area of bus-bar in sq mm/Phase	1800 sq mm
8.	Minimum enclosure size	1150 x 250 mm
9.	Area of enclosure in sq mm	287500 sq mm
10.	Maximum temperature rise during normal operation	40^0C

Calculation for the size of the bus-bar:

Table 5.5

1.	Minimum cross-section area required to withstand short circuit for 1 sec.	$\frac{I}{0.08} X\sqrt{t} = \frac{50}{0.08} X\sqrt{1} =$	625 square mm
2.	De-rating due to material of bus-bar	1.0	—
3.	De-rating due to temperature rise	0.86	Refer to table 5.1
4.	(Bus-bar area)/(Enclosure Area)	(3.5 x 1800)/ 287500 = 2.19%	
5.	De-rating due to enclosure	0.75	Refer to table 5.2
6.	Total de-rating factor	1 x 0.86 x 0.75 = 0.66	
7.	Rating of the chosen bus-bar	2120 amp	(Manufactuer's data)
8.	Thermal rating of the bus-bar	2120 x 0.65 = 1378.0 amp	The required rating is 1150 amp. Therefore the chosen size of the bus bar is appropriate.

5.4 Conclusion

It may be concluded that the bus-duct system is robust if we compare it with the cable. The fault current capacity is more and the voltage drop is less in comparison with cable, but the cost is comparatively more than the cable. Therefore, a trade-off between the cost and other advantages which a bus-duct has over the cable is made.

5.5 Software

Objective

- To calculate the bus-bar size
- To develop specifications
- To store, delete and retrieve the parameters of bus duct from library.

How to use it

- Click on the "Bus-ducts" on the menu of the main screen.
- The Bus-duct screen opens up.
- Fill in the data in the respective boxes and calculate the size of the bus-duct, Refer to Table 5.5 for calculation of size of the bus. Refer to Fig. 5.1 which shows the computer screen for the same calculation done by using the software.

Specifications

Specifications can be developed by clicking on the "Specification" menu on the bus-duct screen. Click "ADD" button and fill in the relevant information in the specified boxes. The specification screen can be printed also by clicking on the Print button. (Refer to Fig. 5.2)

To generate specification in tabular form, click on the "Generate Specification" Button. (Refer to Fig. 5.3)

Library

- The parameters of bus ducts can be stored and deleted in the library. The library screen can be opened by clicking on "Library Entry" on the menu bar (Refer to Fig. 5.2)
- The parameters can be retrieved by clicking on "Library" on the menu bar (Refer to Fig. 5.4)

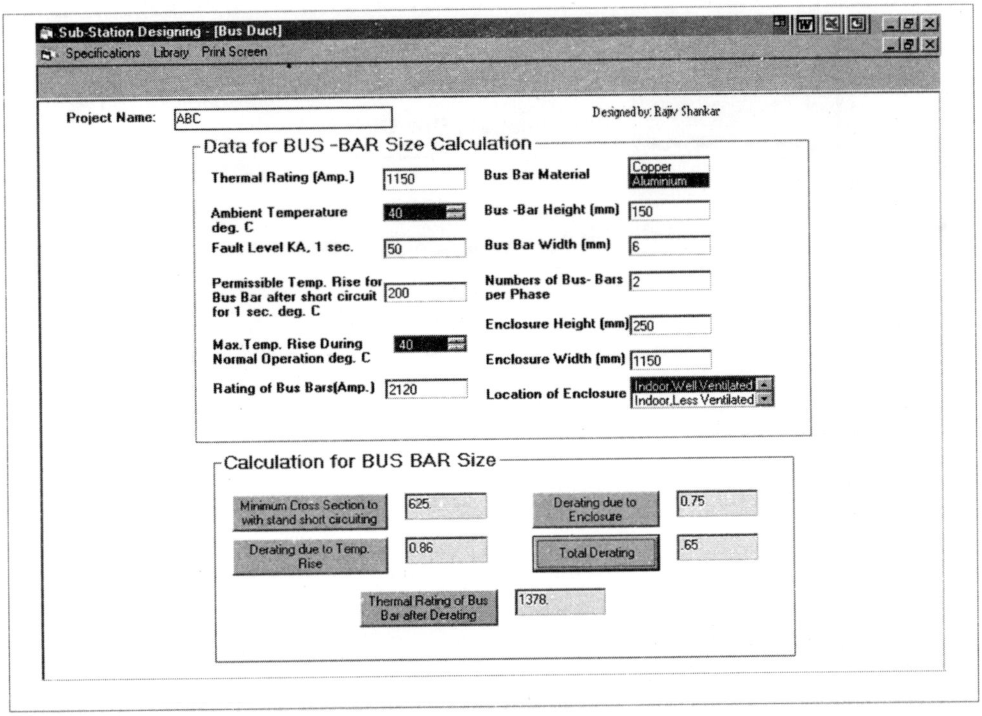

Figure 5.1: Screen for calculation of bus bar size

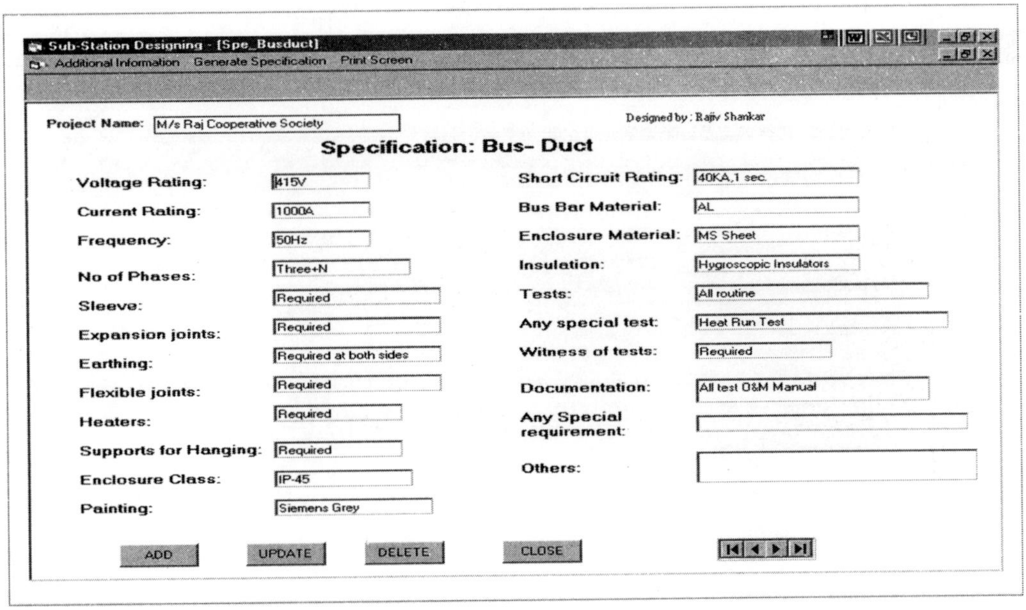

Figure 5.2: Screen for development of specification

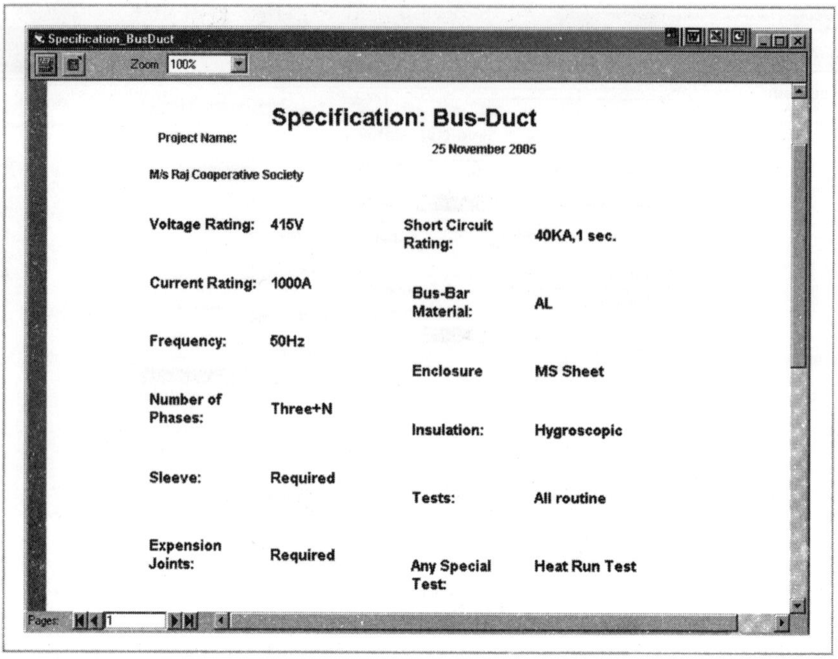

Figure 5.3: Specification in tabular form

Library

The various parameters can be stored, retrieved and deleted in the software by using the same procedure as mentioned in earlier chapters. (Refer to Fig. 5.4 and Fig. 5.5)

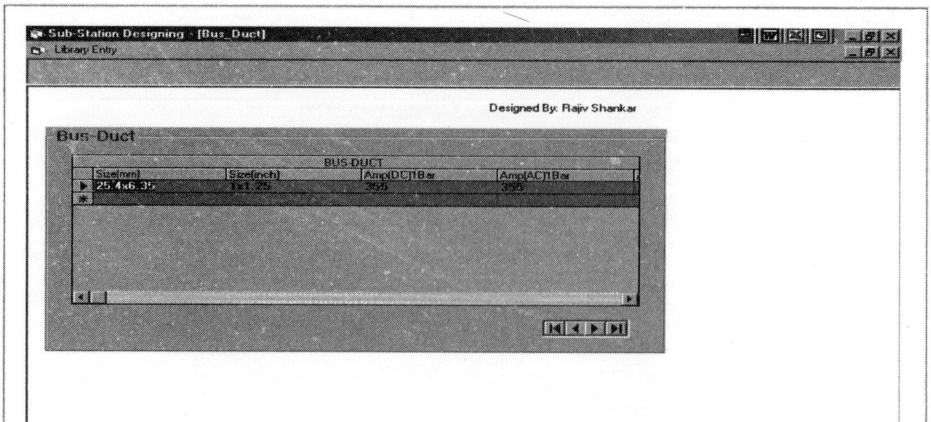

Figure 5.4: Library for bus duct parameters

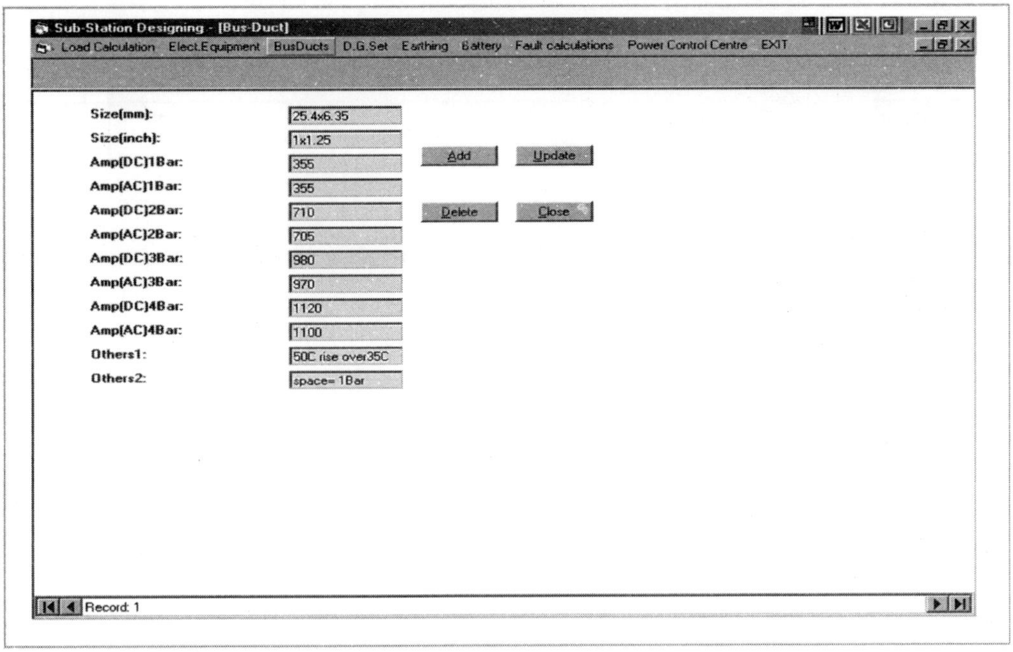

Figure 5.5: Screen for library entry

6 Capacitor-Bank

6.0 Introduction

The capacitor banks are installed in distribution system to compensate for the reactive component of the power, which results in improvement of system power factor, and the main advantage of power factor correction is the reduction in electricity cost. However, the additional benefits are reduction in I^2R losses in the electrical system and improvement in system efficiency, release of additional capacity, reduction in size of electrical equipment for delivery of same amount of power and improvement in voltage regulation.

6.1 Reactive Power

Any system whether it is a large, medium or small plant, commercial establishment, township (large or small) has reactive load. Some of the examples of reactive loads are:

- Motors
- Transformers
- Fluorescent lamps
- Rectifiers
- Furnace
- Various types of electronic equipment
- Computers

These reactive loads draw reactive power from the utility. The current drawn by the electrical equipment has two components. One is active or sometimes called useful component, which is responsible for doing useful work like producing power at a motor shaft. The other component is reactive component, which is also called wattless component. The wattless component is responsible for producing necessary flux across the air gap of an induction motor or in the core of a transformer. It is required for building up flux for magnetic field of an inductive circuit. Without magnetizing current electrical energy cannot flow across the air gap of a motor or in the core of a transformer. The total current as read in the ammeter is the vector sum of these two components. In an inductive load, the reactive component lags at 90^0 from the active component.

This lagging component is responsible for power factor. The power factor is affected by the reactive component. If the magnetizing current (reactive component) is more in comparison to the active component, the power factor will be poor as in a lightly loaded induction motor.

Now, we must address the questions how the poor power factor affects the utility and why the utility companies fix a power factor limit and impose penalty on the consumers if the power factor falls below that limit?

It can be explained in a very simple manner: reactive current will be more for a load of poor power factor, and as mentioned earlier, the reactive current is also a part of the total current supplied by the utility, it loads the equipments of the utility like transmission lines, transformers, switch-gears etc. In another way we can say that for supplying the same useful power to a load of poor power factor more capacity of the equipment is utilized in comparison with a load of good power factor.

6.2 Power Factor

Power factor is the ratio of active power to the apparent power. For example, on an electrical panel ammeter, voltmeter and wattmeter are provided. The ammeter is showing 200 ampere, voltmeter is showing 415 volts and wattmeter is showing 120 KW. Then apparent power will be (1.732 x 200 x 415)/1000 = 143.76 KVA and the power factor will be 120/143.76 = 0.83. Trigonometrically, the power factor is KW/KVA = $\cos\phi$. Generally, any power factor below 0.9 is called poor power factor.

6.3 Remedy for Poor Power Factor

The remedy is to install a capacitor bank or sometimes synchronous motors to improve the power factor. The function of the capacitor banks is to supply the reactive power required locally and relieve the utility from supplying the reactive power to the load. Before installation of capacitor banks we should look into the following points.

- How much compensation is required, to improve the power factor?
- Capacitor bank of what capacity is required?

- Where these capacitor banks are to be located?
- What is the economics of power factor improvement and how it affects the electricity bill?

What should be the power factor?

Many electrical engineers give different answers. Some say it should be 0.9, others say 0.95 and the most common answer is that it should be as near as possible to unity. I think that before fixing a power factor we should look into the following points.

- What is the limit of target power factor imposed by the utility company?
- What is the existing power factor and capacity of system and how much the system is loaded?
- What is the cost of adding additional capacitor banks to improve the power factor beyond the limit as imposed by the utility company. How much additional capacity is released and also its cost?
- Pay-back period of installation of capacitor bank?

If the system has enough spare capacity and additional capacity is not required, the power factor should be kept as per the limit imposed by the utility company, which is normally 0.95, since 0.95 is the most economical power factor. Addition of more capacitor banks may not be economical if additional capacity of the system is not required.

6.4 Location of Capacitor Bank

Now, as far as location of the capacitor bank is concerned, the best location should be as near as possible to the load. Sometimes capacitor banks are installed to act as group correction, i.e. it is connected with the main bus-bar and the main bus has many feeders, since it is not always economical to install the capacitor for individual feeders. For example, if the motors of different rating are installed, many capacitors of different ratings will be needed resulting in increase in cost. Group correction is also economical in cases when different feeders have different loads at different times. Because in such cases automatic power factor sensing relay continuously keeps on sensing the power factor of the load and keeps on switching on and off the capacitor banks to adjust the power factor as per the pre-set value.

Another important point which should be taken care of is that the improvement takes place only from point of application towards the source of power and not in the opposite direction. Therefore, it is always advantageous to place the capacitor bank near the load that reduces the losses in the circuit between the load and the feeder.

6.5 Economics of Power Factor Improvement

For any consumer (except small domestic consumer), the electricity tariff consists of two parts – Demand charges and Unit charges. Demand charges are KVA charges and it depends on power factor. If power factor is low, the KVA will be more and vice versa, and electricity bill for poor power factor will be more. The unit (KWH) charges will not be affected. By installing capacitor bank, the intake of reactive power from the utility grid will get reduced since it is now supplied by the capacitor banks provided locally.

At present, utility companies are not charging for reactive power (KWARH), but it is heard that in near future they are planning to charge for reactive power instead of KVA charges.

Penalty is imposed by the utility company if power factor falls below the specified limit.

Example 6.1

We take a case which will clarify all the points mentioned above. As shown in the single-line diagram (Fig. 6.1) supply is taken at 11 KV and stepped down to 433 V through 11 KV/433V, 1000 KVA, Delta-Star-connected step-down transformer. After stepping down it is supplied to bus-B of power control centre. Bus-B consists of three feeders.

1. Feeder 1 is connected to a motor control centre having a load of 500 KW at 0.85 power factor.The power supply to motor control centre is fed through a 630 sq mm, single core, PVC insulated, aluminum conductor cables. Three numbers of such cables are provided in each phase and two numbers in neutral. The approximate distance between MCC feeder of bus-B and the incomer of bus-C (bus of motor control centre) is 400 metres.
2. Feeder 2 is connected to a miscellaneous load consisting of lighting, lifts, window air-conditioner, computers etc. The total load is 200 KW at 0.75 power factor. It is fed through two numbers, three and a half core, 240 square mm, aluminium conductor, PVC-insulated cables. The distance is approximately 40 metres.
3. Feeder 3 is spare.

The utility has imposed a condition that the power factor should not be less than 0.95. The demand charges are Rs. 250.00 per KVA. Unit charges are Rs. 4.60 per unit.

We would do analysis for working out the KVAR requirement and cost of the capacitor banks for improving the power factor and also for the best location of the bank.

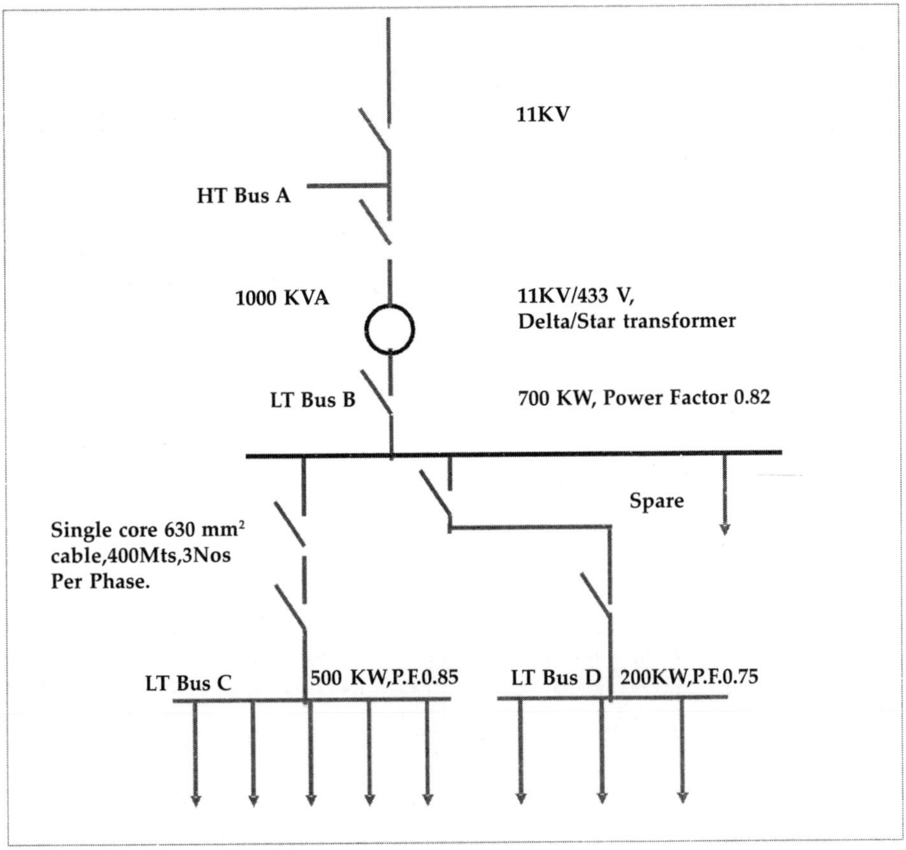

Figure 6.1: Single line diagram of an electrical systems

Analysis

Case 1

Capacitor banks are provided at bus-B (Group compensation)

The total load on bus-B is 700 KW and combined power factor will be 0.82. For improving power factor from 0.82 to 0.95 the KVAR required can be calculated from the following formulae.

$$KW\ (\tan\phi_1 - \tan\phi_2),$$

Where $\cos\phi_1$ is existing power factor and $\cos\phi_2$ is the improved power factor. Alternatively, standard tables (Refer to Table 6.1) are also available to determine the KW multiplying factor for determining the required KVAR for required power factor correction.

The multiplying factor for improving power factor from 0.82 to 0.95 is 0.369.

Therefore the KVAR required is 0.369 x 700 = 258.3

Assuming the cost of supply and installation of capacitor bank per KVAR Rs. 980.00, then the total cost of installation of capacitor banks will be Rs. (258.3 x 980)= Rs. 2,53,134.00

Table 6.1

Existing power factor						Improved			Power		Factor		
	0.8	0.85	0.90	0.91	0.92	0.93	0.94	0.95	0.96	0.97	0.98	0.99	1.00
0.40	1.537	1.668	1.805	1.832	1.861	1.895	1.924	1. 959	1. 998	2.037	2.085	2.146	2.228
0.41	1.474	1.605	1.742	1.769	1. 798	1.83 I	1.860	1.896	1.935	1.973	2.021	2.082	2.225
0.42	1.413	1.544	1.681	1. 709	1.738	1.771	1.800	1.836	1.874	1.913	1.961	2.022	2.164
0.43	1.356	1.487	1.624	1.651	1.680	1.713	1.742	1.778	1.816	1.855	1.903	1.964	2.107
0.44	1.290	1.421	1.558	1.585	1.614	1.647	1.677	1.712	1.751	1.790	1.837	1.899	2.041
0.45	1.230	1.360	1.501	1.532	1.561	1.592	1.626	1.659	1.695	1.737	1.784	1.846	1.988
0.46	1.179	1.309	1.446	1.473	1.502	1.533	1.567	1.600	1.636	1.677	1.725	1.786	1.929
0.47	1.130	1.260	1.397	1.425	1.454	1.485	1.519	1.552	1.588	1.629	1.677	1.758	1.881
0.48	1.076	1.206	1.343	1.370	1. 400	1.430	1.464	1.497	1.534	1.575	1.623	1.684	1.826
0.49	1.030	1.160	1.297	1.326	1.355	1.386	1.420	1.453	1. 489	1.530	1.578	1.639	1.782
0.50	0.982	1.112	1.248	1.276	1.303	1.337	1.369	1.403	1.441	1.481	1.529	1.590	1.732
0.51	0.936	1.066	1.202	1.230	1.257	1.291	1.323	1.357	1.395	1.435	1.483	1.544	1.687
0.52	0.894	1.024	1.160	1.188	1.215	1.149	1.281	1.315	1.353	1.393	1.441	1.502	1.643
0.53	0.850	0.980	1.116	1.144	1.171	1.205	1.237	1.271	1.309	1.349	1.397	1.458'	1.600
0.54	0.809	0.939	1.075	1.103	1. 130	1.164	1.196	1.230	1.268	1.308	1.356	1.417	1.559
0.55	0.769	0.899	1.035	1.063	1.090	1.124	1.136	1.190	1.228	1.268	1.316	1.377	1.519
0.56	0.730	0.860	0.996	1.024	1.051	1.085	1.1 7	1.151	1.189	1.229	1.277	1.338	1.480
0.57	0.692	0.822	0.958	0.986	1.013	1.047	1.079	1.113	1.151	1.919	1.239	1.300	1.442
0.58	0.655	0.785	0.921	0.949	0.976	1.010	1.042	1.076	1.114	1.154	1.202	1.263	1.405
0.59	0.618	0.748	0.884	0.912	0.939	0.973	1.005	1.039	1.077	1.117	1.165	1.226	1.369
0.60	0.584	0.714	0.849	0.878	0.905	0.939	0.971	1.005	1.043	1.083	1.313	1.192	1.333
0.61	0.549	0.679	0.815	0.843	0.870	0.904	0.936	0.970	1.008	1.048	1.096	1.157	1.299
0.62	0.515	0.645	0.781	0.809	0.836	0.870	0.902	0.936	0.974	1.014	1.062	1.123	1.266
0.63	0.483	0.613	0.749	0.777	0.804	0.838	0.870	0.904	0.942	0.982	1.030	1.091	1.233
0.64	0.450	0.580	0.716	0.744	0.771	0.805	0.837	0.871	0.909	0.949	0.997	1.058	1.201
0.65	0.419	0.549	0.685	0.713	0.740	0.774	0.806	0.840	0.878	0.918	0.966	1.027	1.169
0.66	0.388	0.518	0.654	0.682	0.709	0.743	0.775	0.809	0.847	0.887	0.935	0.996	1.138
0.67	0.358	0.488	0.624	0.652	0.679	0.713	0.745	0.779	0.817	0.857	0.905	0.966	1.108
0.68	0.329	0.459	0.595	0.623	0.650	0.684	0.716	0.750	0.788	0.828	0.876	0.937	1.078
0.69	0.299	0.429	0.565	0.593	0.620	0.654	0.686	0.720	0.758	0.798	0.840	0.907	1.049
0.70	0.270	0.400	0.536	0.564	0.591	0.625	0.657	0.691	0.729	0.769	0.811	0.878	1.020
0.71	0.242	0.372	0.508	0.563	0.563	0.597	0.629	0.663	0.701	0.741	0.785	0.850	0.992
0.72	0.213	0.343	0.479	0.507	0.534	0.568	0.600	0.634	0.672	0.712	0.754	0.821	0.964
0.73	0.186	0.316	0.452	0.480	0.507	0.541	0.573	0.607	0.648	0.685	0.727	0.794	0.936
0.74	0.159	0.289	0.425	0.453	0.480	0.514	0.546	0.580	0.618	0.658	0.700	0.766	0.909
0.75	0.132	0.262	0.398	0.426	0.453	0.487	0.519	0.553	0.591	0.631	0.673	0.740	0.882
0.76	0.105	0.235	0.371	0.399	0.426	0.460	0.492	0.526	0.564	0.604	0.652	0.713	0.855
0.77	0.079	0.209	0.345	0.373	0.400	0.434	0.466	0.500	0.538	0.578	0.620	0.685	0.829
0.78	0.053	0.183	0.319	0.347	0.374	0.408	0.440	0.474	0.512	0.552	0.594	0.659	0.802
0.79	0.026	0.156	0.292	0.320	0.347	0.381	0.413	0.447	0.485	0.525	0.567	0.633	0.776

0.80	0.130	0.266	0.294	0.321	0.355	0.387	0.421	0.459	0.499	0.541	0.609	0.750
0.81	0.104	0.240	0.268	0.295	0.329	0.361	0.395	0.433	0.473	0.515	0.581	0.724
0.82	0.078	0.214	0.242	0.269	0.303	0.335	0.369	0.407	0.447	0.489	0.555	0.698
0.83	0.052	0.188	0.216	0.243	0.277	0.309	0.343	0.381	0.421	0.463	0.529	0.672
0.84	0.026	0.162	0.190	0.217	0.251	0.283	0.317	0.355	0.395	0.437	0.503	0.646
0.85		0.136	0.164	0.191	0.225	0.257	0.291	0.329	0.369	0.417	0.477	0.620
0.86		0.109	0.140	0.167	0.198	0.230	0.264	0.301	0.343	0.390	0.450	0.593
0.87		0.083	0.114	0.141	0.172	0.204	0.238	0.275	0.317	0.364	0.424	0.567
0.88		0.054	0.085	0.112	0.143	0.175	0.209	0.246	0.288	0.335	0.397	0.540
0.89		0.028	0.059	0.086	0.117	0.149	0.183	0.230	0.262	0.309	0.369	0.512
0.90			0.031	0.058	0.089	0.121	0.155	0.192	0.234	0.281	0.341	0.484
0.91				0.027	0.058	0.090	0.124	0.161	0.203	0.250	0.313	0.456
0.92					0.027	0.063	0.097	0.134	0.176	0.223	0.283	0.426
0.93						0.032	0.066	0.103	0.145	0.192	0.252	0.395
0.94							0.034	0.071	0.113	0.160	0.220	0.363
0.95								0.037	0.079	0.126	0.186	0.329
0.96									0.042	0.089	0.149	0.292
0.97										0.047	0.108	0.251
0.98											0.060	0.203
0.99												0.143

Pay-back period

KVA at 0.82 power factor will be 700/0.82 = 853.66

KVA charges

853.66 x 250 = Rs. 2,13,415.00

KVA at 0.95 power factor will be 700/0.95 = 736.84

KVA charges

736.84 x 250 = Rs.1,84,210.00

Difference in demand charges Rs. (2,13,415 – 1,84,210) = Rs. 29,205.00

i.e. saving per month = Rs. 29,205.00

Pay-back period will be 2,53,134/29,205 = 8.66 months say approximately eight and a half month.

Additional capacity released = 853.66 – 736.84 = 116.82 KVA

Similarly we can do the calculation for 0.99 power factor improvement.

The KVAR required will be 388.5 and charges for capacitor bank will be Rs. 3,80,730.00. The saving will be Rs. 36,650.00 per month and the pay-back period will be 10.38 months.

Additional capacity released = 853.66 - 707 = 146.66 KVA

Cost of per KVA capacity released when power factor is improved to 0.95 will be 2,53,134/116.84 = Rs. 2,166.50

Cost of per KVA capacity released when power factor is improved to 0.99 will be Rs. 3,80,730 /146.66 = Rs. 2,596.00

Case 2

Capacitor banks are provided at bus-C and bus-D (Individual correction).

Bus-C

The multiplying factor for improving power factor from 0.85 to 0.95 is 0.291.
Therefore the KVAR required is 0.291 x 500 = 145.5
The total cost of installation of capacitor banks will be Rs. (145.5 x 980) = Rs. 1,42,590.00
Saving in demand charges per month = 61.68 x 250 = Rs. 15,421.00

Bus-D

The multiplying factor for improving power factor from 0.75 to 0.95 is 0.553.

Therefore the KVAR required is 0.553 x 200 = 110.6

The total cost of installation of capacitor banks will be Rs. (110.6 x 980) = Rs. 1,08,388.00

Saving in demand charges per month = 56.17 x 250 = Rs. 14,036.00

Total cost = Rs. (1,42,590 + 1,08,388) = Rs. 2,50,978.00

Total saving in demand charges per month = Rs. 29,457.00

Reduction in I^2R losses

Current for 500 KW, three phase, load at 0.82 power factor = 848.32 amp

Resistance of 630 sq. mm cable = 0.047 ohm/kilometre

Resistance of 400 metres length = .0188 ohm

Since three cables are provided in parallel, the total resistance of cable in one phase will be one third i.e. 0.00626 ohm.

The I^2R losses losses in one phase = 4504.99 watt

The losses for three phases = 3 x 4.505 = 13.515 KW

We can summarize the above analysis as following:

- The main benefit in power factor improvement is reduction in electricity bill due to reduction in demand charges
- If the length of feeding cable is more, then we will get the additional benefit of reduction in I^2R losses, and saving in unit charges if the capacitor bank is installed at the load end as seen in the above case study in case of individual feeder correction.
- It is not always economical that the power factor should be near to unity as we have seen that the pay-back period for improving power factor to 0.99 is more than the power factor of 0.95. And also per KVA cost of released capacity is more if power factor is improved to 0.99 instead of 0.95

6.6 Conclusion

1. Finally it is concluded that the main advantage of power factor improvement is reduction in KVA demand, which results in reduced power cost. In addition, the losses in the system get reduced and a saving in energy is realized. But as compared to saving which is obtained in KVA charges, it is very small and installation of capacitor bank just for saving due to reduction in losses is not justified. It should always be looked at as additional benefit.
2. Lastly, one more important point to add – before installation of capacitor bank measurement of harmonic in the system shall be made, and if it is more than the permissible limit, the harmonic filter shall be provided since presence of harmonics adversely affects the performance of capacitor bank. The stability of stand-by/captive power generators may be affected if power factor is improved to unity and excitation current becomes zero.

6.7 Software

Objective

- Calculation of KVAR requirement for a given load and power factor.
- Calculation of reduction in losses.
- Development of specifications for capacitor bank.

How to use it

- Click on the "Electrical Equipments" and then on "Capacitor Bank" on the main menu (Refer to Fig. 6.2)
- Capacitor bank screen opens up and the various calculations like KVAR requirement, reduction in losses, economics in selection of capacitor banks can be made by filling in the required data in the respective boxes and clicking the appropriate button. (Refer to Fig. 6.3)

  Refer to Example 6.1. In this example, the calculated KVAR requirement will be 258.3 for providing a capacitor bank on Bus B.

The load on transformer is 853.65 KVA at 0.82 p.f. and I^2R losses at rated load are 12.5 KW. The I^2R losses at the 853.65 KVA load will be 9.03 KW and the losses in cable 13.515 KW.

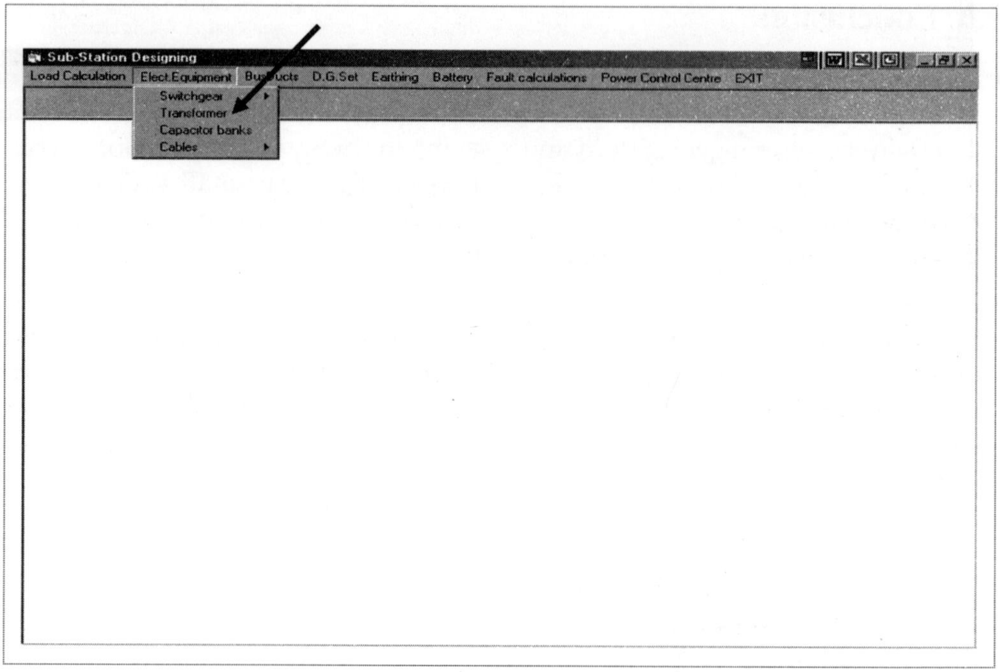

Figure 6.2: Main screen

$$\left(\frac{85365}{1000}\right)^2 \text{ x } 12.5$$

$= (0.72)^2 \text{ x } 12.5 = 9.03\text{KW}$

Therefore, the total losses will be (9.03 + 13.515)KW = 22.545 KW. The reduction in losses by improving power factor will be, Losses (KW) x (1-(origional p.f.i.e. 0.82)^2/ (improved p.f. i.e. 0.95)^2) = 5.75 KW.

Reduction in unit charges/year = 5.75 x 8760 x 4.6 = Rs. 231702 (Assuming 8760 loading hours)

Reduction in demand charges per month = Rs. 29205.00 (as calculated in Example 6.1)

Yearly reduction in demand charges = Rs. 350460.00

Total saving in electricity bill per year = Rs. 582162.00

Cost of installation of capacitor bank = Rs. 253134 (As calculated in Example 6.1)

Pay-back period = 253134/582162 = 0.4348 year

Or 5.21 months. Refer to Fig. 6.3, which shows the computer screen for the above calculation.

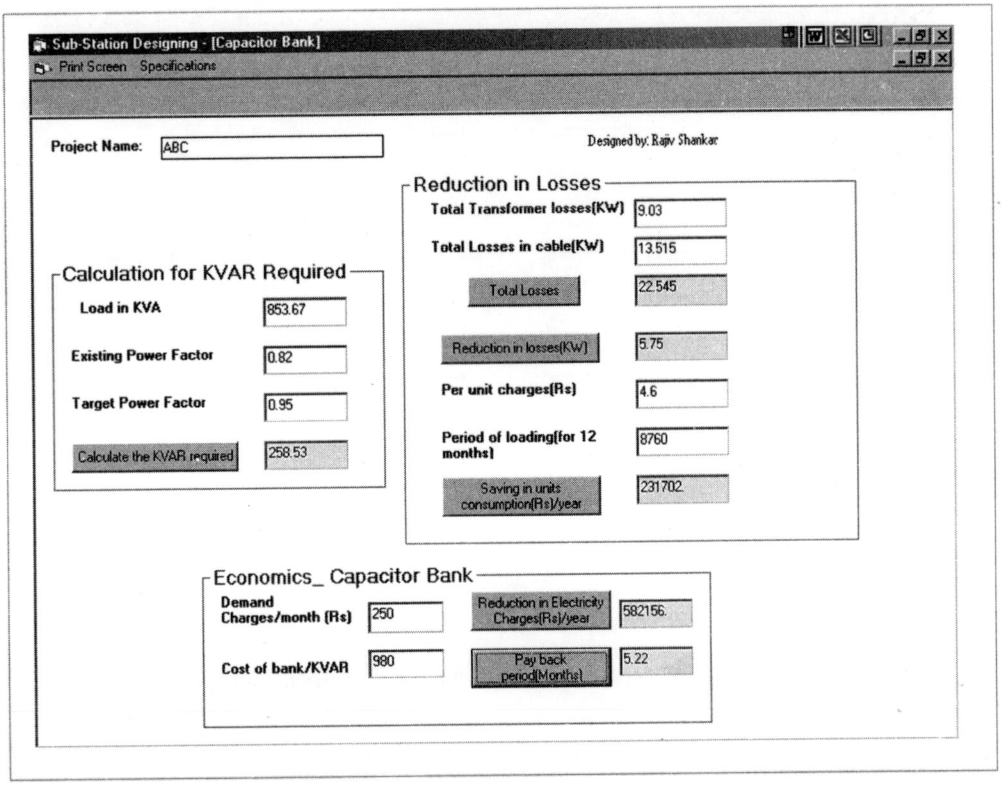

Figure 6.3: Screen for calculation of KVAR requirement and reduction in losses

Specification

- Specifications can be generated by clicking on "Specification" on the menu, and all the required parameters are to be filled. Refer to Fig. 6.4.
- Click on "Generate Specification" for generating specification in a tabular from (Fig. 6.5)

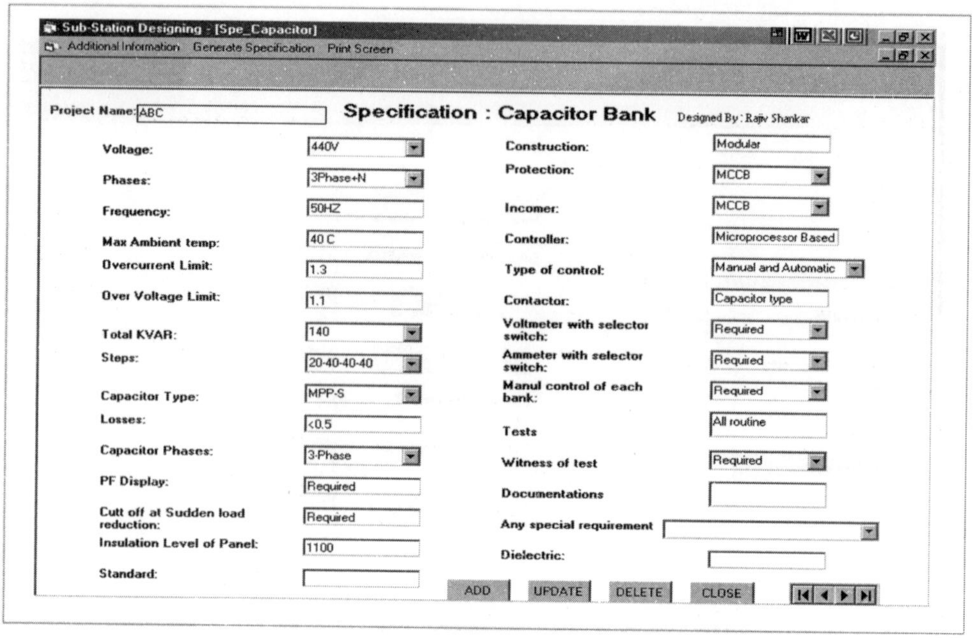

Figure 6.4: Screen for development of specification

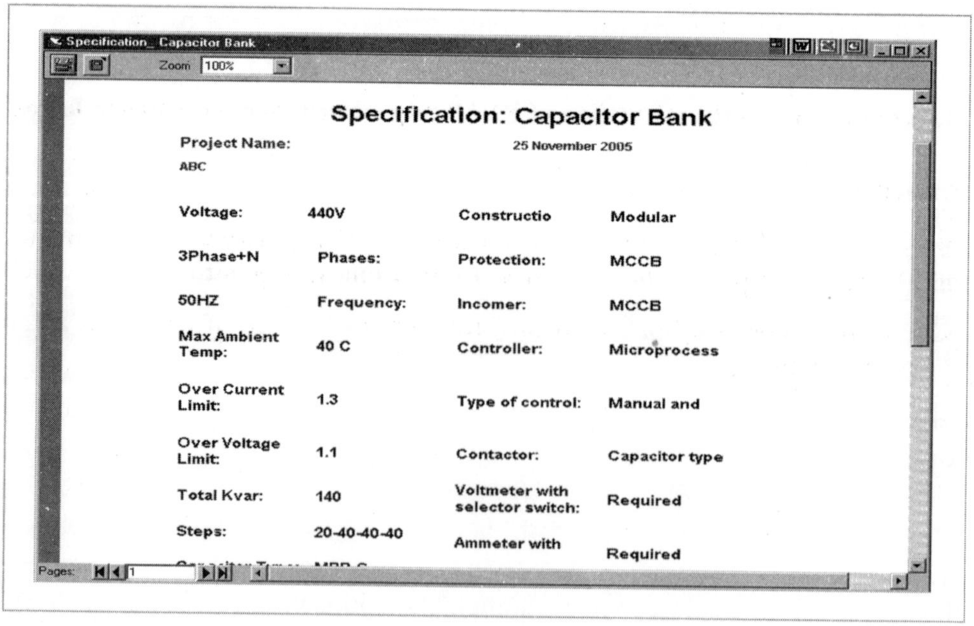

Figure 6.5: Screen for generation of specification in tabular form

7 Grounding

7.0 Introduction

Grounding or earthing is one of the essential components of any electrical system. It plays a very important role as far as the safety of the electrical equipments and also of human beings is concerned. Basically two types of grounding are provided.

- System grounding
- Equipment grounding

Firstly, let us discuss system grounding – what is system grounding and why is it required? In a three-phase power system, the system grounding is grounding of neutral of the equipments connected in a system like grounding of neutral of transformer or generator. Earlier, grounding of neutral was not done and neutral became floating. The system was not so complex in those days as it is now. Some people argue that continuity of the system is better in an ungrounded system. There is no doubt that the continuity is better because the tripping does not take place due to ground fault in one phase out of three phases, and tripping will only take place when the other phase also develops the ground fault. There are some serious disadvantages, or in more appropriate terms, serious hazards in an ungrounded system.

- In an ungrounded system full line-to-line voltage appears throughout the system on occurrence of ground fault on one line. This voltage is 1.732 times higher than the line to neutral voltage. The insulation can withstand such stress, but frequent or prolonged stress will cause deterioration and reduction in life of insulation. Sometimes multiple failure of equipments takes place.
- The second ground fault will result in high current equivalent to line-to-line fault current to flow between the two faults and lead to tripping of one or both the circuits.

- When a system having line-to-ground fault is switched on, high transient over voltages may appear and the repeated re-striking of arc during interruption of line to ground fault may result in failure of equipment in other circuits.

There are many advantages in grounding the system as listed below:

- The over-voltage is suppressed and the insulation of the equipments are not subjected to high voltage resulting in the increased life of the equipments.
- The protective relaying system can be deployed to clear the earth fault fast and selectively.
- The system becomes safe.
- The equipments can be protected against over-voltage from other reasons like falling of a high voltage conductor on low voltage conductor, striking of lightning etc.
- In case of ground fault, high ground current can be prevented to flow through the ground by providing resistive grounding. This results in prevention of damage of equipment due to burning of the faulted point.

7.1 Types of System Grounding

- Solid grounded
- Resistance grounded
- Reactance grounded
- Transformer grounding

Each type of grounding system has its own advantages and disadvantages. The engineering personnel decide which type of grounding is more suitable for a particular system. They choose the grounding system after looking into different other aspects like safety of equipment and human safety, relaying system etc. In general, for low voltage system solid grounding and for medium and high voltage system resistance grounding is more prevalent. The arcing fault is a high-resistance fault. It can take place between the two phases in an ungrounded system and between phase and ground in a grounded system. The normal over-current relays cannot detect the fault and if detected, it cannot clear the fault quickly because the current flowing through the circuit will not be appreciable. The arcing fault releases a large amount of energy and burning of equipment takes place. The heat may vaporize the conductor and the surrounding metal enclosure may also get damaged. Sometimes the damage is so severe that the entire equipment needs to be replaced. The best way to prevent such fault is to provide a relay in the neutral grounding, which can monitor and clear the arcing fault at the earliest.

The system grounding is obtained by grounding the neutral of the source like transformer or generator. Where the neutral is not available at the source, a grounding transformer, like zig-zag transformer or star-delta transformer, is used.

7.1.1 Solid grounding

The low voltage system is solidly grounded, since high ground fault current is required to operate the protective devices. If the source of power supply is transformer only, then ground fault current may be equal to three-phase fault current. However, if power source is generator then the ground fault current should be less than the three phase fault current to prevent damage of the slot and slot winding where the fault has taken place. To achieve this, a low value resistance or a reactor is used in the neutral of the generator. If the generator is capable of tolerating the ground fault current equal to the three-phase fault current, then solid grounding is preferred. The reason for solid grounding is that it is cheap, and LT switch gear's tripping coil connected in series requires a large value of ground fault current. Damage from high value of ground fault current may be prevented because the protective device is fast enough to trip the system immediately. Further, if ground fault is protected through a circuit breaker and relay arrangement, then on operation of the relay all the three phases open, while if it is protected through the fuses, then only the fuse of the effected phase blows, and the other two fuses may not blow, and the condition of single phasing may take place.

In low voltage system where solid grounding system is provided, residual relaying scheme is more common. As shown in Fig. 7.1 three current transformers are provided. The protection for short circuit and over-current is provided through the relays provided in phase of the star-connected secondary winding of the current transformers while the ground fault protection is provided by residually connected ground fault relay. Under normal condition there would be no residual current flowing in the residual path of the secondary of the current transformer, but if the ground fault takes place, residual current flows and the ground fault relay picks up. The limitation in such a system is that if the resistance of the ground fault is large, for example, falling of an overhead line on the bushes, the ground fault current will not be large enough to operate the relay. The minimum pick-up current will also depend on the current transformers ratio.

As mentioned above, if the earth fault current is low enough, residually connected relay may not be able to operate. The modern digital or numerical relay with low setting can operate on low earth fault current if low setting is done, but the older electro-mechanical relay may have difficulty in operation due to high burden on current transformers.

For such a situation, a core balance current transformer may be the solution. As shown in Fig. 7.2, it is mounted around all the phases and neutral (if present). The current in the secondary winding of the transformer will be proportional to the residual current The medium voltage transmission lines are usually solidly grounded, since grounded neutral also serves as the surge arresters. Generators are usually not connected directly with the transmission lines.

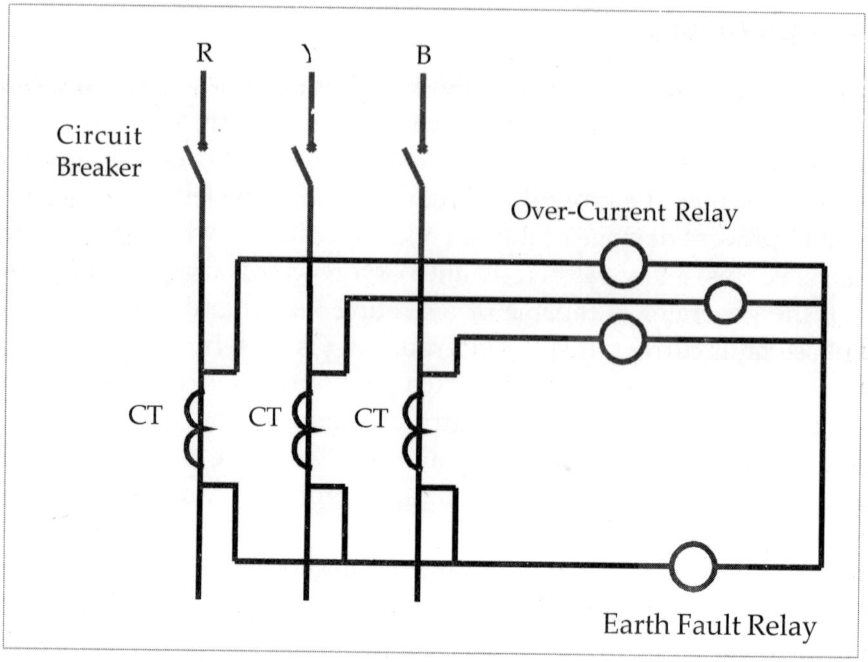

Figure 7.1: Residual relaying scheme

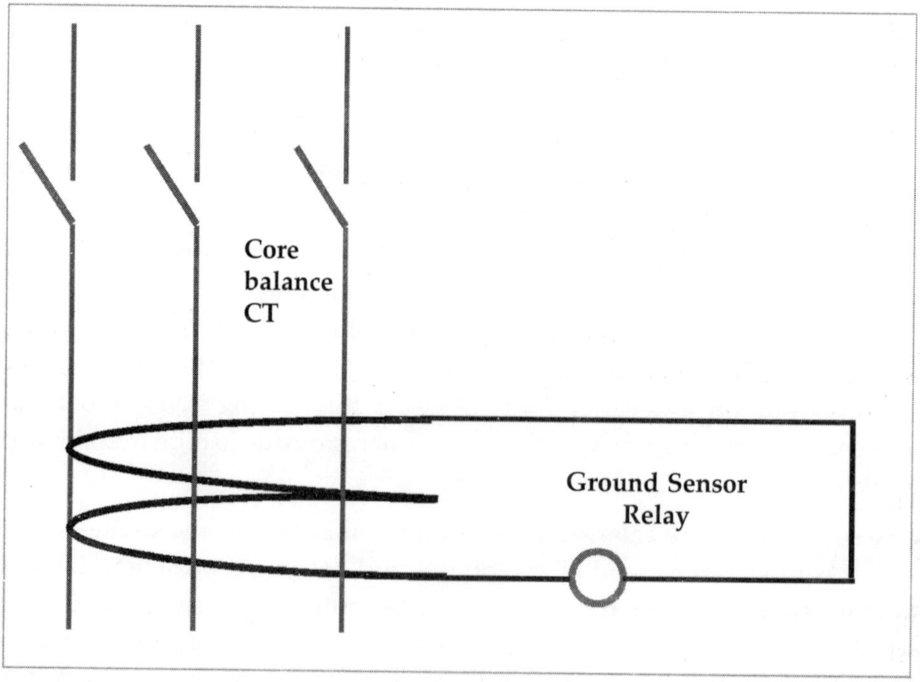

Figure 7.2: Ground fault protection by using core balance current transformer

7.1.2 Resistance grounding

As per IEEE-142, the transient over-voltages may be prevented if the resistance of the neutral grounding does not exceed the limit of $R_0 >= 2X_0$. And the ratio of zero sequence resistance to positive sequence reactance is positive and less than one.
In medium voltage system resistance grounding method is adopted since the ground fault current is large and the rotating equipment like alternator may get damaged.

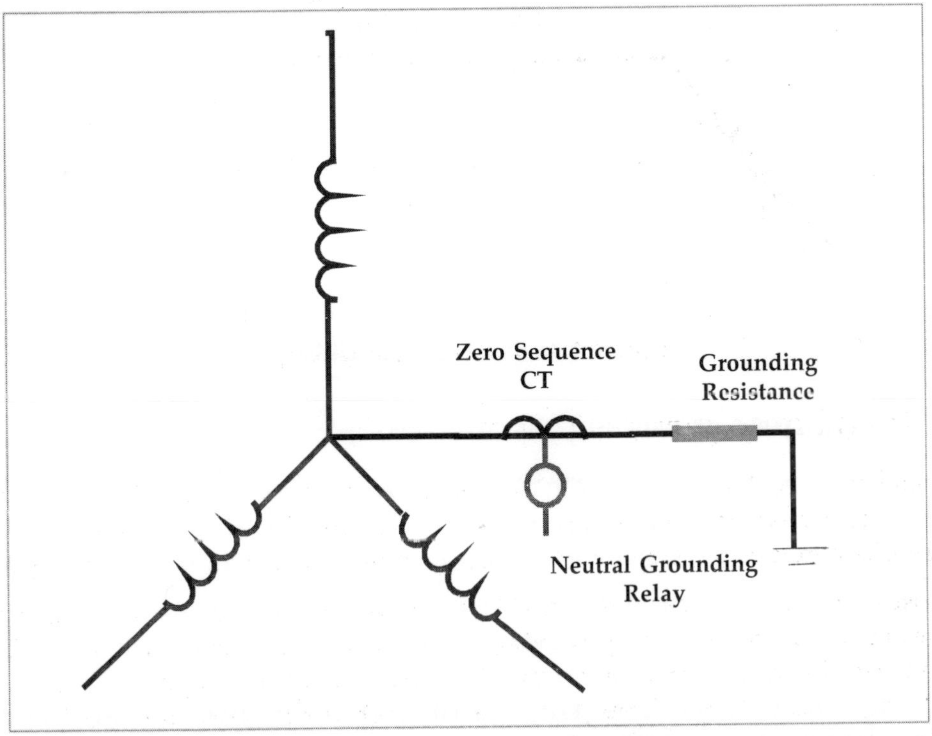

Figure 7.3: Resistance grounding

In high and medium voltage system where neutral is grounded by a low resistance to limit the ground fault current, ground fault protection is done by providing a zero sequence current transformer in the neutral. A time delay over-current or instantaneous ground fault relay is provided in the secondary of current transformer (Fig. 7.3).

7.1.3 Reactance grounding

Reactance grounding is done by connecting a reactor between neutral and the ground (Fig. 7.4). The purpose is to limit the ground fault current. The reactance grounding is used where the generator neutral is connected to earth. The reactor is added in the generator neutral to limit the ground fault current to a value less than the three-phase fault current.

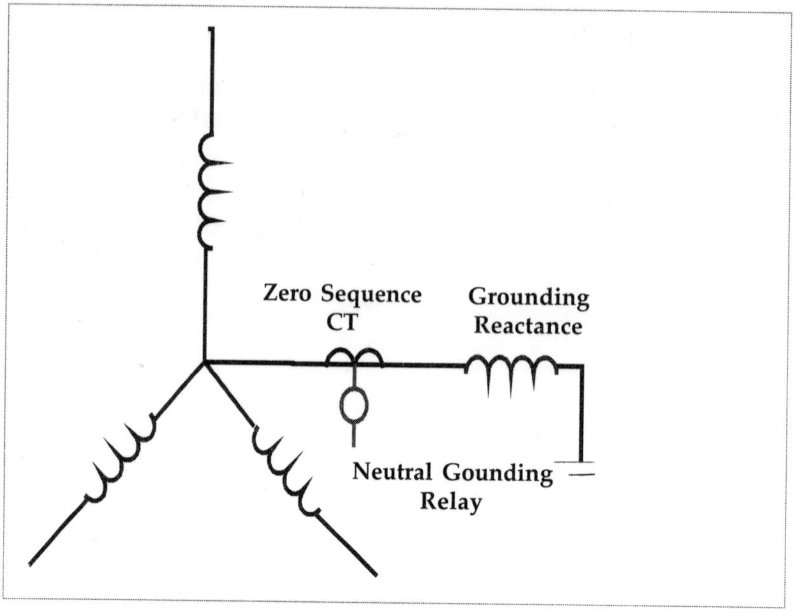

Figure 7.4: Reactance grounding

7.1.4 Transformer grounding

The grounding transformer is used in old system where system neutral is not available or in the existing system where the delta system is required to be grounded. Zig-zag or Wye-delta connected transformer is used to obtain the neutral of the system.

The most common type of grounding transformer is three phase zig-zag type of transformer. It has no secondary winding. The connection is shown in Fig. 7.5. The impedance of the transformer for normal three-phase current is high and only a small magnetizing current flows through the transformer during normal condition, but it provides a low impedance path to the zero sequence current flowing through neutral during ground fault in the system. The ground fault current flows into three windings of the transformer and the windings are provided in such a manner that the current in one winding on one leg flows in opposite direction to the current of another winding provided on the same leg. This results in equal division of the ground fault current in the three lines and low impedance from transformer to ground.

Another method is to provide Wye-delta connected three phase transformer and the neutral of the Wye is connected to ground.

The rating of the grounding transformer in KVA should be equal to line to neutral voltage times the rated neutral current. The resistance grounding is normally provided to earth the neutral of the grounding transformer. The grounding transformers are designed to carry the rated current for a short duration, for example 10 seconds to 1 minute. Therefore, the size of grounding transformer is much smaller than the size of the continuously rated transformer of the same rating.

The grounding transformer is either directly connected to the system main bus through the circuit breaker or it is connected between the circuit breaker and the main transformer. If it is connected with the main transformer, then grounding transformer is required for each transformer and the protection of the grounding transformer is included with the main transformer. The grounding transformer is required to be connected with earth all the time.

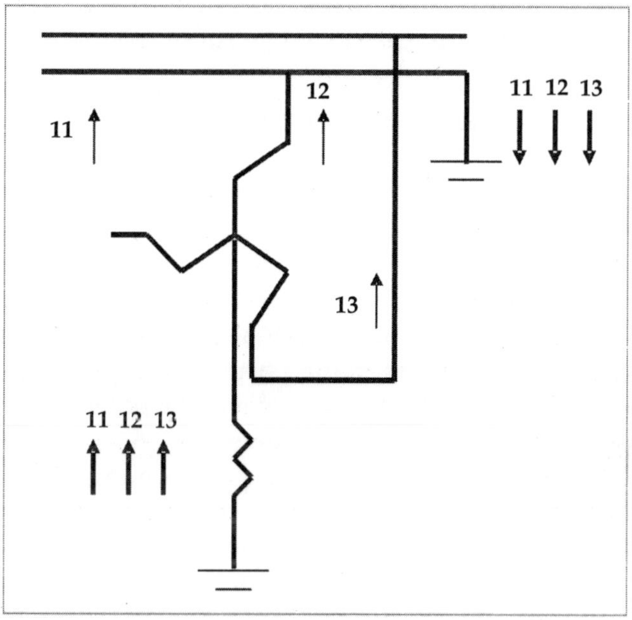

Figure 7.5: Zig-zag type neutral grounding transformer

7.2 System Grounding Points

Grounding shall be done at each voltage level. The neutral point available at transformer, generator shall be grounded. If neutral point is not available, then grounding transformer should be used. This grounding transformer should be connected at the source end. Grounding should be done always at source end. It is not advisable to ground the neutral at load end. The reason is: if grounding is done at load end, it is difficult to ensure that the load will always remain connected. If the load is not in service, then the grounding will also be disconnected, and it is difficult to make special arrangement to ensure that the neutral of load should always be connected to ground whether it is in service or not.

When there are many sources like transformer, generators of different ratings, it is desirable to ground them individually. The ground fault current will increase if we combine the neutral of all the sources. Another way is to make a common bus of the neutral of all the similar types of sources and ground them through the resistance. The value of resistance shall be chosen such that the practical minimum current flows through the resistance and the value of the current will not increase beyond a certain value, it will help in better relaying system. Sometimes when two or more than two alternations are put in parallel, the practice of connecting, only one neutral is adopted and the neutral of other sources is kept floating. The reason is that circulation of third harmonic current and zero sequence current is prevented. However, in resistance grounding system the harmonic current will be minimum and the neutral of all the sources can be connected to the system. This practice makes the system simple. The manufacturer of the equipment like alternator shall also be consulted regarding connection of neutral of all the alternators while running in parallel.

7.3 Calculation of Ground Fault Current

When designing the grounding system, it is essential to calculate the grounding current.

Now the question arises that if the source has more than one transformer and generator, then how many transformers and generators should be grounded, and whether they should be grounded separately or all the neutral points of all the sources should be combined to form a bus. Generally, in industrial premises it is difficult to keep the grounding of the neutral and the metallic part in isolation so that at the time of occurrence of ground fault in one particular system, the potential of the other healthy system does not rise. Therefore, for a large system it is preferred to form a mesh-type grid system. This mesh is connected with earth pit at a number of places, and all the equipments are connected with this mesh either directly (LT system) or through resistance in the neutral (Medium or high voltage system).

7.4 Equipment Grounding

The basic purpose of equipment grounding is human safety, by preventing the rise of potential to any dangerous level, which is unsafe and can cause shock to any human being when accidentally he may come in contact of the metallic enclosure of a live conductor. By grounding the metallic enclosure, it is kept at ground potential. The grounding conductor should be such that the earth fault current could safely pass to the ground without excessive temperature rise or without any spark/arc, which can cause fire.

If the loop impedance is not low enough as in case of arcing fault, the protective device may not operate. In such a situation the residual-current-operated protective device should be used.

Proper coordination with the upstream and downstream relays should be done; otherwise opening of main supply breaker will result in the loss of power supply for a large section and loss of production.

Metal work like fences, pipes, sheath of cables, which are not associated with the power system shall also be grounded and connected with the common grid.

A detailed calculation shall be done for the size of ground conductors, earth electrodes etc. This calculation is normally done through the empirical formulas, which are given in any standard code.

It is a statutory requirement and the electrical system cannot be energized until and unless proper grounding is provided. Various requirements related to the grounding system have been specified in the IS: 3043 and Indian Electricity Rule .

Some of the provisions specified in the Indian Electricity Rule have been mentioned below.

- Neutral conductor of the three-phase four-wire system and the middle conductor of two phase 3 wire system shall be grounded by no less than two separate and distinct connections with minimum of two different ground electrodes or such large number as may be necessary to bring the earth resistance to a satisfactory low value, both at generating station and at distribution station.
- In the case of system comprising electric supply lines having concentric electric cable, the external conductor of such cable shall be grounded by two separate and distinct ground connection with earth.
- The connection with ground may include a link by means of which the connection may be temporarily disconnected for the purpose of testing or for locating a fault.
- In case of an alternating current systems, there shall not be inserted any impedance, cutout or circuit breaker in the connection with ground.
- Alternating current system, which are connected with earth as mentioned before, may be electrically interconnected.
- The frame of every generator, stationary motor, portable motor and metallic part of all transformers and any other apparatus used for regulating and controlling energy and all medium voltage energy consuming apparatus shall be grounded by two separate and distinct connections with earth.
- All metal casing and metallic covering protecting any electric supply line and apparatus shall be connected with ground and shall be so joined and connected across all junction boxes and other openings as to make good mechanical and electrical connection throughout the whole length.
- When the supply of energy is at low voltage, the tube light, fans etc. shall be grounded at one point. All plug sockets of three-pin type shall be grounded at third pin permanently and efficiently.
- All grounding system shall consist of equi-potential bonding conductor capable of carrying the prospective earth fault current and groups of pipe/rod/plate electrodes for dissipating the current to the general mass of ground without exceeding the allowable temperature limits as per IS 3043 in order to maintain all non-current-carrying metal works reasonably at ground potential and to avoid dangerous contact potential being developed on such metal work.

- Ground resistance shall be sufficiently low to permit adequate fault current for the operation of protective devices in time and to reduce neutral shifting.
- It should be mechanically strong, able to withstand corrosion and retain electrical continuity during the life of the installation. All grounding systems shall be tested to ensure adequate grounding before the electric supply lines or apparatus are energized.
- All grounding system belonging to the supplier shall in addition be tested for resistance on dry day during the dry season not less than once every two years.
- A record of every grounding test made and the result must be kept by the supplier for the period of not less than two years after the day of testing, and shall be made available to the inspector.
- The supply of energy to every electrical installation other than low voltage installation below 5 KW shall be controlled by earth leakage protective device so as to disconnect the supply instantly on the occurrence of earth fault and leakage current

All non-current-carrying metal parts associated with HV/EHV installation shall be efficiently earthed to a grounding system or mat which will

- Limit the touch and step potential to a tolerable value.
- Limit the ground potential rise to tolerable values so as to prevent danger due to transfer of potential through ground earth wires cable sheath, fences, pipe lines etc.

7.5 Design of Earth Electrodes

The ground resistance of an electrode consists of

- Resistance of the electrode itself.
- Contact resistance between the electrode and the soil.
- Resistance of the soil from the electrode surface outward in geometry set-up for the flow of current outward from the electrode.

The resistance of the soil depends on the resistivity of the soil, which vary area to area. The soil resistivity can be reduced by treating the soil, i.e by adding salt, charcoal and water.

Ground electrodes

Generally, the following types of ground electrodes are provided in a small/medium level of installation:

- Plate
- Pipe or rod
- Strip or conductor

Empirical formulas are given in IS 3043 for calculation of resistance to ground of each type of electrode mentioned above. These formulas are reproduced below.

7.5.1 Plate electrode

$R = \rho/4 \sqrt{\pi/A}$
ρ = resistivity of soil (assumed uniform) ohm-metre
A = area of plate electrodes of both sides (Square metres)
R = Approximate resistance to earth of a plate
The standard size of plate is 600 cm x 600 cm, the thickness should be as following:
Cast Iron electrode: should not be less than 12 mm
GI electrode: should not be less than 6.3 mm
Copper electrode: should not be less than 3.15 mm

7.5.2 Pipe electrode

$R = (100 \times \rho \times \log_e 4l/d)/(2 \times \pi \times l)$
l = length of rod or pipe (cm)
d = diameter of rod or pipe (cm)
ρ = resistivity of soil (ohm-metre)
R = Approximate resistance to earth of a pipe or rod
Cast iron pipe may be 2.5 to 3 metre long with 100 mm diameter and 13 mm thick, GI and MS pipe may be 38 to 50 mm diameter.

7.5.3 Strip or conductor electrode

$$R = \left(\frac{100\rho}{2\pi l}\right) \log_e \left(\frac{2l^2}{wt}\right)$$

ρ = resistivity of soil (ohm-metre)
l = length of strip in cm
w = depth of burial of electrode in cm
t = width (cm) of the strip
It is further mentioned in the IS 3043 that if the resistance of single plate is higher than the required value, two or more plates may be used in parallel and the total resistance will be inversely proportional to the number employed, provided each electrode is installed outside the resistance area of the other i.e. two electrodes shall have separation of minimum 2 metres. Electrodes are normally connected through a grounding strip to form a grid. IS 3043 does not specify any formula for calculating the total resistance of grid. It just says that the total resistance will be inversely proportional to the number of electrodes. Approximate calculation of grid resistance is done by dividing the resistance obtained from the above formula by the number of electrodes provided. It is desirable that the total resistance of the grid should be less than one ohm. However,

it is also important that sufficient earth fault current should flow through the system so that the protection system could operate properly.

Another factor which should also be taken into account is the current loading of the electrode so that the earth fault current could be dissipated safely and without any temperature rise.

As per IS 3043, the maximum permissible current density for plate-type electrode is as given below:

$I = 7.57 \times 10\ ^{8}/\sqrt{(\rho \times t)}$ Amp/ sq.metre

t = duration of the earth fault (sec.)

ρ = resistivity of the soil (ohm-metre)

The earth potential rises during the fault condition. This may cause damage to the telecommunication cable if it falls in that area; further, cattle or human beings may also get shock. Therefore it is also necessary to calculate the step and touch potential. The step and touch potential can be calculated from the following formulae. IEEE Std.80 (1986)

$E\ step = (6 \times \rho s + 1000) \times 0.116/\sqrt{t}$

$E\ touch = (1.5 \times \rho s + 1000) \times 0.116/\sqrt{t}$

ρs = resistivity of soil (ohm-metre)

t = duration of shock current (sec.)

As mentioned earlier, ground fault current is very low in case of overhead line and may take a longer time to clear the fault. To keep the step and touch potential low, it is desirable to earth the neutral conductor at a number of points after some fixed interval by a pipe or rod electrode.

Cross-section area of grounding conductor

The cross-section area of the grounding conductor shall be calculated from the following formula

$I/S = K(1/\sqrt{t})$

S = Cross-section area (square mm)

t = Operating time of disconnecting device (sec.)

K = Factor dependenting on the material of the protective conductor

I = Ground fault current (amp.)

Value of K

Material	Copper	Aluminium	Steel
1 sec. current rating in A/ sq mm	205	126	80
3 sec. current rating in A/sq mm	118	73	46

7.6 Conclusion

The general aspects of system grounding and equipment grounding have been covered in this chapter. Formulas are available in many standards for calculating the size and numbers of grounding electrodes, which mainly depends on the resistivity of earth. Often soil treatment is also required to reduce the resistivity. Computer programs are also available for more complex calculations for a large electrical system. The grounding system should be designed in such a way that it fulfils the two requirements.

1. Human safety

2. Quick isolation of faulty equipment

Sometimes it becomes very difficult to properly coordinate the ground fault relay, since the magnitude of ground fault current varies depending on the location of the fault and sufficient ground fault current should flow to isolate the system quickly. However, it is desirable that the total resistance of the ground grid should be less than 1ohm.

7.7 Software

Objective

The objective of the software supplied with this book is to calculate the numbers of ground electrodes for a specific size and also to develop the specifications of the ground electrode.

How to use it

- Click on the "Earthing" on the main menu.
- A screen opens up. Then click on the respective Tab, i.e. "Pipe Electrode", Plate Electrode" or "Strip electrode" for calculation of number of electrodes.

Example 7.1

Let's take an example for calculating the numbers of earth pits (Plate type) with the following available data.

Soil resistivity: 11 ohm-metre
Plate size: 600 x 600 mm
$R = \rho/4 \sqrt{(\pi/A)}$

ρ = resistivity of soil (assumed uniform) ohm-metre
A = area of plate electrodes of both sides (Sq metres)
Plate Resistance to earth
R = 11/4 √ 3.14/0.72 = 5.74 ohm

If one earth electrode is provided, then the plate resistance to earth will be 5.74, the required resistance is 1 ohm. As mentioned in the standard IS 3043, if more than one electrode is provided in parallel and outside the resistance area of each other, and are connected together, then the total resistance will be inversely propotional to the number employed.

Then approximately, 5 earth electrodes are required to achieve the combined grid resistance of 1 ohm. This is only the approximate calculation. Refer to Fig. 7.6 which shows the computer screen for this calculation.

Example 7.2

Pipe Electrode:
Soil resistivity: 11 ohm-metre
Diameter of pipe: 40 cm

Length of pipe: 150 cm

$R = (100 \text{ x}\rho \text{ x } \log_e 4l/d)/2 \text{ x } \pi \text{ x } l$ ohm

l = length of rod or pipe (cm)

d = diameter of rod or pipe (cm)

ρ = resistivity of soil (ohm-metre)

$R = (100 \text{ x } \rho \text{ x } \log_e 4l/d)/(2 \text{ x } \pi \text{ x } l)$ ohm

$= (100 \text{ x } 11 \text{ x } \log_e 4 \text{ x } 150/40)/(2 \text{ x } \pi \text{ x } 150)$ ohm

= 3.0 ohm

As mentioned earlier, if one ohm resistance is required then approximately three earth electrodes are required. Calculation has also been done through software in Fig. 7.7.

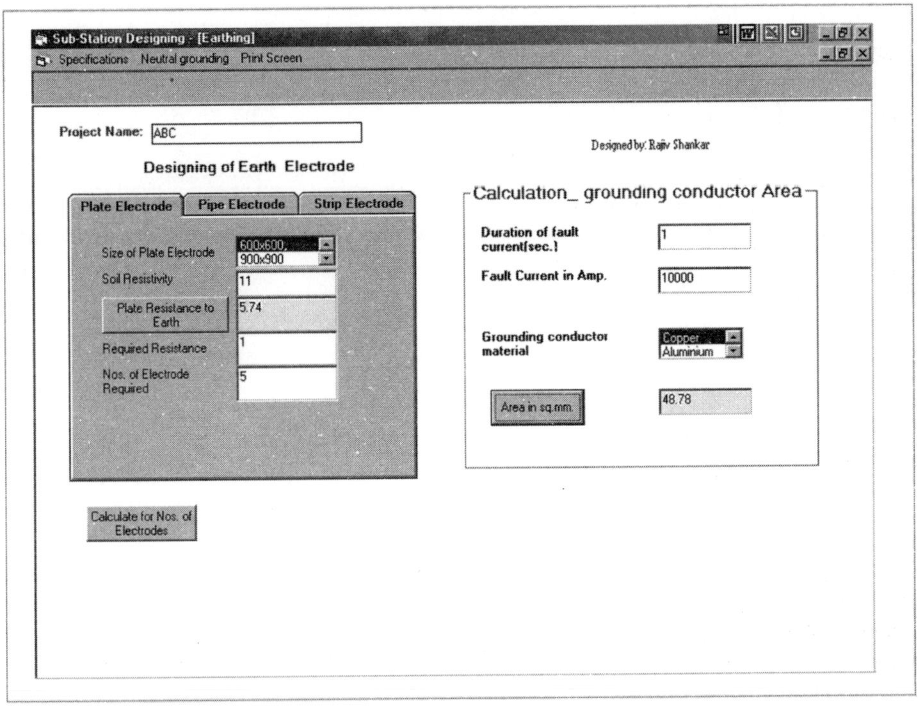

Figure 7.6: Screen for calculation of resistance to earth of plate type electrode

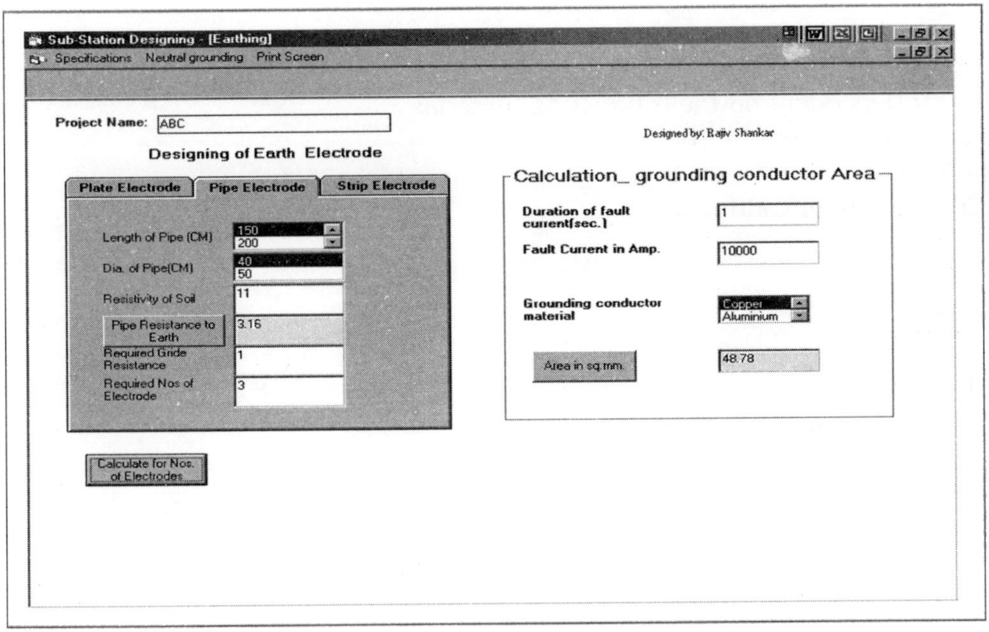

Figure 7.7: Screen for calculation of resistance to earth of pipe type electrode

Example 7.3

Strip Electrode:

$$R = \left(\frac{100\,\rho}{2\,\pi l}\right) \log_e \left(\frac{2l^2}{wt}\right)$$

R = resistivity of soil (ohm-metre)
l = length of strip in cm
w = depth of burial in cm
t = width (cm)
Data:
l = 1500 cm
w = 75 cm
t = 5 (cm)
ρ = 11 (ohm-metre)

$$R = \left(\frac{100 \times 11}{2 \times 3.14 \times 1500}\right) \log_e \left(\frac{2 \times 1500^2}{75 \times 5}\right)$$

= 1.096 ohm
Calculation done through software is shown in Fig. 7.8.

Example 7.4

Cross-section area of grounding conductor:
$I/S = k(1/\sqrt{t})$
S = cross section area (sq mm)
t = operating time of disconnecting device (sec.)
k= factor dependent the conducting material
Ground fault current I = 10 KA
Time = 1 sec.
K= 205 for copper conductor

$$S = \frac{I\sqrt{t}}{k}$$

$$= \frac{10{,}000\sqrt{1}}{205}$$

S = 48.78 sq mm

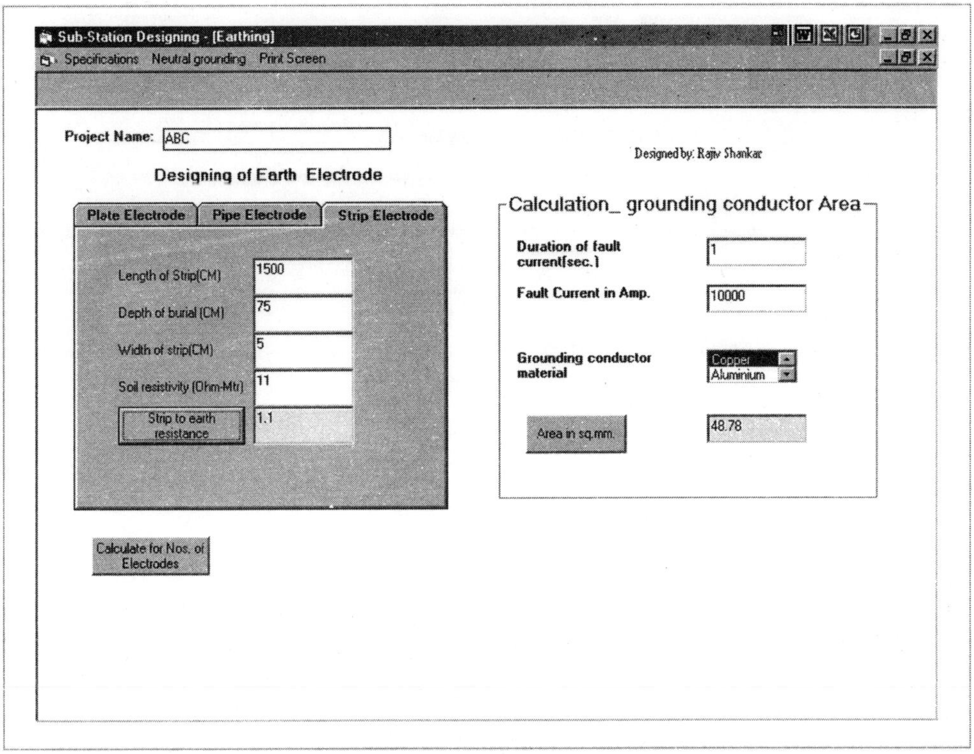

Figure 7.8: Screen for calculation of resistance to earth of strip type electrode

Specifications

Specifications of the earth electrodes can also be developed by clicking on the "Specification" on the menu and filling in the boxes.

Refer to Fig. 7.9 and 7.10, which show the computer screen for the specifications. The procedure of generation of specification is the same as mentioned in earlier chapters.

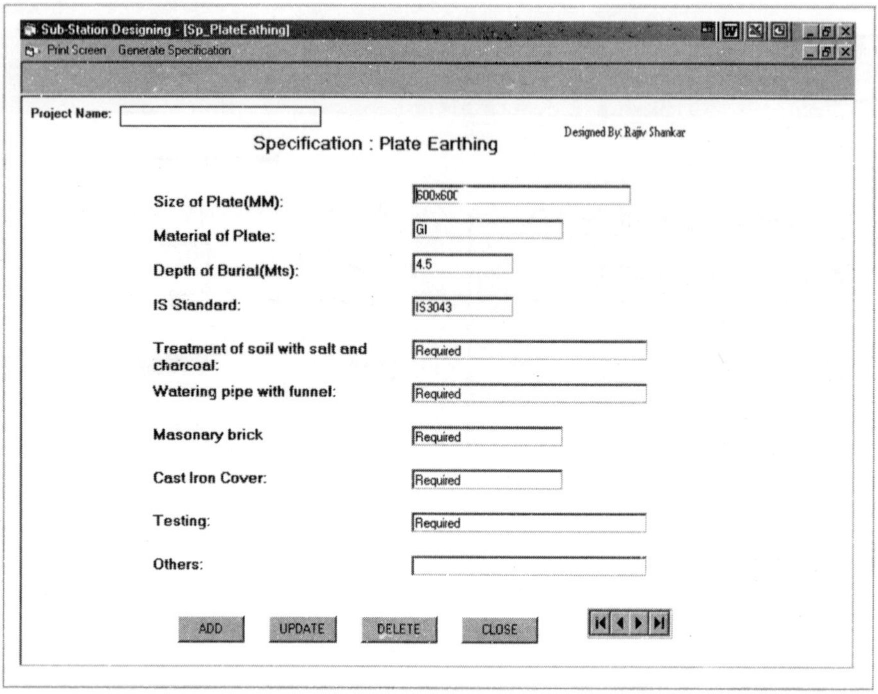

Figure 7.9: Screen for developed specification of plate earthing

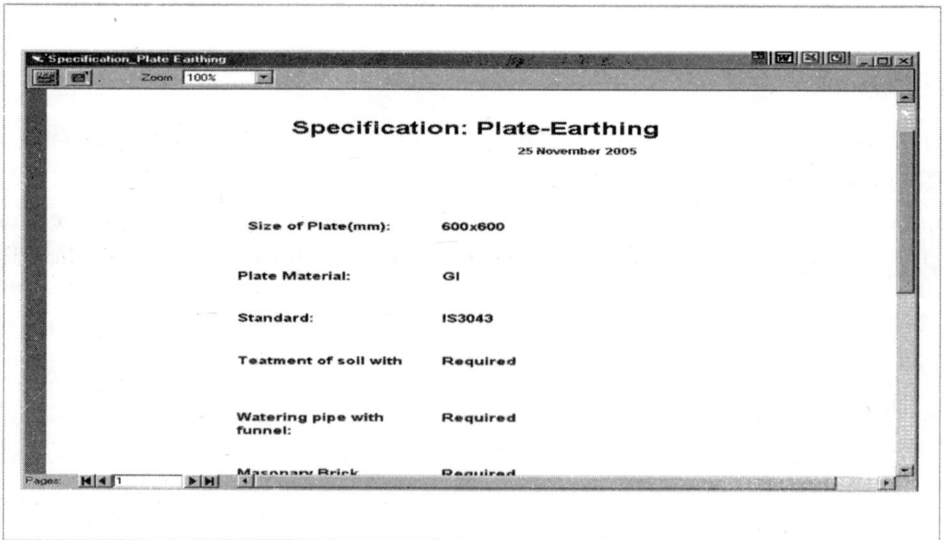

Figure 7.10: Screen for generation of specification in tabular form

8

Battery Bank and Battery Charger

8.0 Introduction

The batteries are an integral part of switchgear installed in a substation. Many types of batteries are available. However, we will keep our scope restricted to lead-acid, and nickel-cadmium batteries, since these batteries are generally used for switchgear control and protection schemes. Besides this, these batteries are also used for starting of DG sets, if DG Sets are installed for supplying power to emergency loads. Two types of batteries are available depending on the construction of the batteries.

1. Wet-type battery.

2. Maintenance free-valve-regulated battery.

Two types of maintenance-free batteries are available for construction.

Absorbed Glass Mat (AGM) and the Gel type.

It can be further classified as Starting (Cranking) or deep or high discharge, and deep cycle batteries. The high discharge batteries are used for cranking the engines and are designed to deliver a quick burst of energy. The plates are thinner and there are more plates. The deep cycle battery has instant energy but greater long-term energy delivery. The plates are thicker and can survive a number of discharge cycles.

8.1 Lead-Acid Battery

The lead-acid battery consists of a negative plate of pure lead and a positive plate of lead oxide. The sulphuric acid (35% sulphuric acid and 65% water) is used as electrolyte. A separator is used between the positive and negative plates. The specific gravity of the electrolyte of a fully charged battery should be 1250 at 27^{0}C. The electrons are released due to electro-chemical reaction in the battery. The basic electro-chemical equation is as follows:

$$PbO_2 + Pb + 2\,H_2SO_4 \underset{\text{Discharge}}{\overset{\text{Charge}}{\rightleftharpoons}} 2\,PbSO_4 + 2\,H_2O$$

During the discharge of batteries lead and lead dioxide react with sulfuric acid and lead sulfate; water and energy are created. The reverse takes place during the charge cycle. The lead sulfate and water are converted to lead, lead dioxide and sulfuric acid.

The lead acid batteries are available in three basic designs of positive plate:

- Pasted plate
- Tubular plate
- Plante plate

Pasted plate

It consists of a metallic grid of lead antimony or lead-calcium alloy. The openings in the grid are filled with paste of active material, i.e. lead-oxide paste. Lead-antimony is used for the application where temperature is more and frequent cycling is required. Lead-calcium is used where interval is longer between cycles.

Tubular plate

As evident from the name, porous tubes are provided on the lead antimony grid and powdered active mat rial i.e. lead-oxide is filled into the tubes. The tubes are sealed from top as well from bottom. The advantage of this type of design is that the cycling capability increases and the power density also increases, i.e. it provides more ampere-hour capacity per cubic foot of battery volume.

Plante plate

The positive plate is cast with complex grooves or with circular openings. The lead oxide is provided in thin layers. The life of plante plate battery is longer in comparison to other type of lead-acid battery.

The negative plate is made of pasted plate type in all the above mentioned types of batteries.

8.2 Nickel-Cadmium Battery

The positive plate of each cell consists of nickel hydroxide and the negative plate consists of cadmium hydroxide. The positive and negative plates immersed in solution of potassium hydroxide in de-mineralized water or battery-grade water with lithium hydroxide as an additive. The lithium hydroxide increases the life of the battery. The electrochemical reaction takes place as below:

$$\text{Charge} \qquad\qquad \text{Discharge}$$

$$2\,Ni(OH)_3 + Cd \rightleftharpoons 2\,Ni(OH)_2 + Cd(OH)_2$$

On charging the cadmium hydroxide is reduced to cadmium, and nickel hydroxide attains a higher level of oxidation. The process is reversed on discharge. The electrolyte takes no part in the chemical reaction and just behaves like a carrier of ions.

Pocket plate type is the most common type of battery in nickel-cadnium category. This can be further classified as low discharge, medium discharge and high discharge type. Low discharge type can carry a load above 2 hours and up to 20 hours. It is used in fire alarm system, emergency lighting, telecommunication, photovoltic and cathodic protection. The medium discharge type can carry current above 40 minutes and upto 3 hours. The application is switchgear protection, instrumentation and process control, UPS, emergency and train lighting. The high discharge type can carry current upto one hour and normally used in diesel generating set starting, UPS, air/helicopter ground starting, electromagnets etc.

The construction of positive and negative plates of battery are same. The active material is sandwiched between two perforated nickel plated steel strip. These strips are crimped together to form a long narrow pocket. Several pockets are joined mechanically and cut in suitable length to form plate pieces. Both the cut edges of the plate pieces are sealed in the U section steel strip by spot welding. The plate assemble is compressed together to have a contact between the active material, perforated steel strip and U section.

8.3 Selection

The selection of battery depends on the type of application for which it is required. For example, the battery for engine starting will be different from the type of battery which is required for switchgear application. Discharge rate, backup time, operating voltage, duty cycle environmental conditions etc. are the other factors which influence the selection of battery.

The nominal voltage of a fully charged lead-acid battery is 2 volt per cell, while it is 1.2 volt per cell in case of Ni-Cd battery. Therefore, more cells of Ni-Cd are required in comparison to lead-acid for the same application.

Battery voltage gradually declines during the discharge and manufacturer specifies the end cell voltage. The voltage should not go below the end cell voltage during discharge and a cut-off relay is normally provided to cut off the load if the battery voltage goes beyond the end cell voltage. This end cell voltage is also important for selection of the battery because this will be the minimum limit of the voltage of the load across which batteries are connected. The rate of voltage decline depends on the current demand of the load, duration of discharge, design and type of cell, age and temperature of cell. The manufacturer specifies the discharge rate of the battery.

8.4 Environment

The battery life depends on the environment. Normally, batteries are designed for wide range in temperature fluctuation; however, beyond that limit the life of the battery is affected. Actual capacity is the function of ambient temperature. The capacity increases when the temperature rises above 27°C and decreases as the temperature falls. It is estimated that the battery life reduces considerably if rise in ambient temperature takes place. Further, if the batteries are installed at such a place where regular maintenance is not possible, then maintenance-free batteries should be used. A very good charger is required for charging the maintenance-free battery-bank because if the battery is charged at more than the recommended current, the excess gas will be released through the vent and recombination will not take place, resulting in damage of batteries.

8.5 Comparison of Wet and Sealed Low-Maintenance Battery

Table 8.1 shows a comparison between sealed and wet batteries.

Table 8.1

Sl. No.	Characteristic	Sealed battery	Wet battery
1.	Safety	Safe, no spillage of acid, no special room is required, explosion proof, leak proof.	Special room with ventilation and acid proof floor tiles is required. Explosion has been observed in batteries due toexcessive gassing.
2.	Space	Minimum space is required.	More space is required.
3.	Topping up	Because of recombination of oxygen and hydrogen gases in the battery, no water topping is required.	Water topping is required since water is lost during charging of the battery.
4.	Life	Long life, if the surrounding temperature is kept at the specified level and the batteries are charged at proper charging rate. High charging current may cause excessive generation of gasses which may be released through the valve and drying up	Short life. Some manufacturers claim 6 to 8 years float life for tubular batteries at 27°C temperature.

		of the battery will take place. Some battery manufacturers claim 20 years float life at 27°C temperature.	
5.	Corrosion	The grid is made of special calcium alloy and corrosion is less.	Grid is made of high antimony alloy and corrosion is more.
6.	Container	Light and compact cell container is used since the batteries are not very bulky and can be installed in horizontal position also. The separator is (AGM) absorbent Glass Mat	To hold a large quantity of electrolyte heavy-duty and bulky container (hard rubber) is normally used. The separators are made of polyvinyl chloride. Batteries are always kept in vertical position. Terminal corrosion may take place.
7.	Performance	Low internal resistance, better discharge characteristics.	High internal resistance, poor discharge characteristics.
8	Transportation	Can be transported in filled and charged condition and ready to use condition.	Cannot be transported in filled and charged condition. Electrolyte has to be carried and filled at site and initial charging is required at site.
9	Operation cost	Negligible	Operation cost is more. Regular maintenance and topping up of batteries are required.
10	Self Discharge	Low self discharge	High self-discharge

8.6 Sizing of the Battery

The following steps are taken in selecting the capacity of the battery:

8.6.1 Rate of discharge

Since the capacity of the battery varies if the discharge time is different from the specified. Normally 10 hours (lead-acid) or 5 hours (Ni-Cd) discharge capacity is specified at 27°C temperature to an end voltage of 1.75 V/cell in case of lead-acid cell and 1.0 V/cell in case of nickel-cadmium cell. For example, if a cell is rated for 300 amp hour capacity at a 3 hours discharge rate, the cell should be able to deliver 100 amp current for three hours but the same cell will not be able to deliver 300 amp for one hour because if the discharge rate is increased, the A-H capacity reduces, while on the other hand if the discharge rate decreases, the cell will be able to deliver current for much longer time. The rate of discharged factor is the ratio of capacity performance at any discharge rate to performance at rated capacity. This can be calculated by dividing discharge time in hours by a K-factor for that end cell voltage and given time. The K-factor is specified by the manufacturer.

It is relatively simple to calculate the A-H capacity of the battery if current discharge is constant (Refer to Example 8.1). However, in switchgear application the load current is not constant and also cyclic in nature. It is important to prepare a discharge profile of the battery. The discharge profile should indicate the current, duration and sequence of each load. The load on the battery can be divided into three categories:

- Momentary current required for closing, tripping of switchgear.
- Continuous current for indicating lamp, holding coil, supervisory control equipment.
- Emergency light and other power equipment.

In normal condition the batteries are kept in float condition and the DC load is supplied by the charger; but if the normal power supply fails, the battery should be capable to supply the continuous load and also the momentary load like operation of circuit breaker etc. In normal case it is rare that entire bank of circuit breakers are closed or opened at a same time, which will impose a higher load on the battery. Further, the timing of momentary load is also important. For example, battery is supplying a continuous load to emergency light and other equipment and it has reached the end of the cycle, and at that time it is required to close or open one or more circuit breakers; the battery should be capable of delivering that momentary load.

8.6.2 Temperature correction

The performance of a battery depends on the temperature, i.e. capacity increases or decreases with the rise and fall in the temperature. The manufacturer specifies the capacity at 27^0C temperature.

8.6.3 Ageing factor

The performance of a battery comes down with time. Therefore, to deliver its rated capacity even at the end of estimated life, ageing factor is taken into account. Normally, a correction factor of (1.25) is taken as ageing factor.

8.6.4 End cell voltage

The acceptance of minimum voltage of a cell on discharge is the limiting factor for the deliverable capacity of the battery.

Example 8.1

Battery is required to be selected for a constant discharge application and the following information is available.

- Load current 10 amp for three hours.
- End cell voltage 1.75 V/cell
- System voltage 24 V

Numbers of cells: 24/2 = 12 nos.

Minimum Voltage with 12 nos of cell = 12 X 1.75 = 21 V

Capacity = 10 X 3=30 AH

Discharge factor = say 0.75

Ageing factor = 0.8

Capacity = 30/(0.75 X 0.8) = 50 AH

8.7 Comparison between Lead-Acid and Nickel-Cadmium Battery

Table 8.2

Sl. No.	Characteristic	Nickel-cadmium battery	Lead-acid battery
1.	Cost	Costly in comparison to lead acid. Maintenance free.	Less costly. Maintenance cost is more.
2.	Storage	Can be stored for a long time in partial charged (40%) state. Cycling (charge-discharge) is needed if the battery is stored for more than six months to regain the full capacity.	Cannot be stored in partially charged stage. It will result in sulfation of battery. Sulfation will cause reduction in capacity or sometimes sudden death if it is in advance stage. Therefore, the battery is always kept on float charging.
3.	Charging	Fast charging is possible in case of nickel cadmium battery. In fact nickel-cadmium batteries prefer fast	Fast charging is not possible in lead-acid batteries. Multi-stage charger is preferred. In stage 1 constant current is applied raising

		charging. Fastest full charge time may be approximately over one hour. Fast charging improves the efficiency of the battery. The battery should not be left in the charger for more than a few days even if on trickle charge because of memory effect.	the cell voltage to preset voltage. The current is limited to the maximum charging current as specified by the manufacturer. It takes approximately 5 hrs and battery is charged to 70%. In stage 2 current is reduced. Full charge is reached after the battery voltage has reached its threshold. The final stage is float charge, which compensates for the self-discharge of the battery. Correct setting of voltage limit is very essential. Overcharging would lead to grid corrosion on the positive plate, while undercharging will cause sulfation of the battery. High voltage (2.5 V/cell) will also cause gassing and loss of electrolyte. The battery will progressively heat up due to overcharging. As it gets hotter, it will accept more current and heat up further. This is called thermal runaway and can destroy the battery in as little time as a few hours.
4.	Discharge	The nickel cadmium battery should be discharged to full cycle to prevent the memory effect once in every one to three months.	Full cycle discharge should be avoided. Up to 80% depth of discharge is recommended.
5.	Temperature	Charging a hot battery decreases the charge time. The battery may not get fully charged.	Warm temperature lowers the battery voltage. Overcharge may occur if the cut off voltage is not reached. Optimum operating temperature is 25^{0}C.
6.	Life	Long life, can provide over 1000 charge-discharge cycle.	300 to 500 charge-discharge cycle.
7.	Self discharge	40% three months	40% per year

8.8 Memory Effect

The term memory is derived from the myth that nickel-cadmium battery can remember how much energy was drawn in the preceding discharge. If the battery is left in the charger for some days for quite some time or repeatedly charged without periodic full discharge, then the voltage and capacity will drop. Periodic discharge up to 1 volt per cell is essential to prevent the memory effect of the battery.

The reason is that active cadmium material is present in finely divided crystals. These crystals are small in a good battery, and therefore the surface area is more. But in due course of time if the battery is left in the charger for a long time without discharging it completely, the crystals grow and conceal the active material from the electrolyte. In the advance stage the sharp edges of crystals penetrate the separators causing high self-discharge or sometimes short circuiting of the battery.

If the battery is not discharged fully as mentioned above, it becomes difficult to restore it. For reconditioning, the battery is discharged below the threshold limit (1 V/Cell). The current is kept low to prevent cell reversal. If the battery is not discharged for 6 to 12 months, it becomes difficult to restore it, and such a battery should be replaced.

The nickel-cadmium battery should not be left in the charger for more than a few days; not even on trickle charge. The batteries should be discharged completely (1V/Cell) once in every two months. The batteries should not be discharged before each charging to avoid undue stress on the battery. The temperature should not rise more than the specified limit during charging.

Nickel-cadmium battery cannot be disposed easily since it contained toxic material.

The self-discharge in nickel-cadmium batteries is comparatively more. It is approximately 10% of its capacity per month. Further, as the temperature increases, the self-discharge also increases. The rate of self-discharge doubles with every 10^0C increase in temperature. The self-discharge also increases after a few hundred cycles. The reason for this is that the battery plate begins to swell and presses against the separator.

The cell capacities cannot be predicted accurately even with modern manufacturing techniques. The cell in a bank should be within ±2.5% . The reason is that the capacity of a weak cell is less and is discharged more quickly in comparison to other cells. This may result in cell imbalance and cell reversal on weak cell if discharged too low. On charge, the weak cell charges first and goes into heat-generating overcharge, while the healthy cell will go on accepting charge. The weak cell becomes further weak and the cell mismatch increases further.

8.9 Sulfation

The biggest problem in lead-acid battery is sulfation due to chronic under-charging. The capacity of a cell is lost due to sulfation. The sulfate ions enter into deep bonds with the lead on the cell's plate. The sulfate ions can bond with the lead at three successively deeper energy levels. Level 1 is the bond when the batteries are charged and discharged. After a month or so, at level 1 some of the bonds form level 2 bonds. Level 2 bonds require more energy to break the bonds. After several months, at level 2 the ions form level 3. No amount of recharging will be able to break the level 3 bond. Equalizing charges should be tried to restore the battery. The equalization charge is a controlled overcharging of an already fully recharged battery. During equalization charges the cell voltage will become very high, about 2.7 V/Cell. This overcharge contains necessary power to break the level-two bonds and force them to level one. Once they reach level one, the bond is easily broken and the sulfate ions re-enter into electrolyte solution.

The lead-acid cell has low energy density, poor weight to energy limits. It cannot be stored in a discharged condition i.e the cell voltage should never be dropped below 2.1 V. Normally, lead-acid battery is well suited for stand-by applications that require-occasional deep discharge.

8.10 Specifications of the Battery

Table 8.3 shows specifications of lead-acid and nickel-cadmium batteries.

Table 8.3

Sl. No.	Type	Nickel-cadmium	Lead-acid
1.	Construction	Wet pocket plate, valve regulated pocket plate (maintenance free)	Tubular battery, plante plate battery, valve regulated Lead-acid (maintenance free)
2.	Discharge	High discharge, medium discharge Low discharge. The type of battery depends on the application. For example, for cranking of engine, high discharge batteries are required, while for switchgear etc. medium or low discharge batteries are required. In general, Ni-Cd batteries are being manufactured for 5 hours discharge rating.	Similar to nickel-cadmium batteries, the lead-acid batteries can also be categorized as high discharge medium discharge low discharge. In general. lead-acid batteries are being manufactured for 10 hours discharge rating.
3.	Container	Translucent polypropylene.	Translucent polypropylene, Transparent.
4.	Nominal capacity	To be specified depending on the application.	To be specified depending on the application.

5.	Recommended charging current limit	As per manufacturer's specification. Normally it is 0.2 C, where C is the battery capacity.	As specified by the manufacturer.
6.	Recommended float charge voltage	1.4V to 1.42 V/Cell	2.17 to 2.25 V/Cell
7.	Recommended boost voltage	1.54 to 1.69 V/Cell	2.2 to 2.5 V per cell
8.	Cell dimensions	Usually specified by the manufacturer.	Usually specified by the manufacturer
9.	Accessories	Nickel plated cell Connectors, nut bolts and other hardware, racks for keeping the cells	Lead coated cell connectors, Nut bolts and other hardware, racks etc.
10.	Conditions of the cells at delivery	Charged, partially charged or discharged. Batteries are supplied in charged condition.	Charged or discharged. Normally, wet batteries are supplied in discharged conditions and they are charged at site after installation. The maintenance-free batteries are supplied in charged conditions.

8.11 Battery Charger

The performance of battery charger greatly affects the life of the battery. The reason for this is that overcharging and undercharging both are harmful for battery and affects the performance to a great extent. If the charging voltage is high, excessive current will flow into the battery and after reaching full charge, it will cause decomposition of water in electrolyte and hence it results in premature aging. At high rate of overcharge, a battery will progressively heat up. As it gets hotter, it will accept more current, heating up even further. This is thermal runaway and can destroy the battery within hours.

If low charge voltage is applied, the current flow will stop before the battery is fully charged. It will cause some of the lead sulfate to remain on the electrodes, which will reduce the capacity.

It is very important to ask the manufacturer the maximum charging current for the battery, and the charger shall have in-built current limiting device so that the charging current should not exceed the specified value of current setting.

Constant current/constant voltage charging method is the best method of charging. The charger is also called float-cum-boost charger. In many applications, the charger remains connected with the load and performs the dual functions i.e. it float- or boost-charges the batteries and also supplies the load current. When the batteries are in discharged state and require to be charged quickly, the boost mode of charging is adopted. In this mode the voltage setting is kept high and current-limiting device is set such that the maximum current does not exceed the maximum charging current as specified by the manufacturer. When charging is nearly complete, the voltage rises to a specified value and the current decreases, the charger switches to lower setting of voltage, which is called float mode. In this float mode the lost charge due to self-discharge of the batteries is replenished. Normally, a good battery charger performs the boost-and-float charging function automatically when the selector switch is kept on auto operation. Since the load also remains connected with the charger, care should be taken that the load is not subjected to high charger output voltage when the charger is in boost mode. To prevent this, diode droppers are provided in the charger, which are automatically connected between the charger and load to drop some of the voltage across it.

8.12 Specifications of Battery Charger

Generally, the following parameters are specified while preparing the technical enquiry of battery charger.

- Type of charger: Float charger, boost charger, float-cum-boost charger
- Diode dropper: Required or not required
- AC input voltage: Three phase, 415 V (Voltage variation is also specified) or single phase with neutral
- AC input frequency: 50 Hz with variation ±3%
- DC output voltage: Nominal
- Float and Boost voltages
- Output current limit
- Power conversion: Full wave controlled bridge rectifier/half controlled bridge rectifier
- Power Devices: Transistors, thyristors
- Efficiency: To be specified by the manufacturer
- Ripple voltage: Not more than 3% RMS when battery is connected
- Switchgear and Protection

  Input: Fuse or MCB

  Output: Fuse or MCB

Current limiter Circuit

Soft start feature

Surge suppressor

Control circuit fuses

Semiconductor fuses in bridge circuit

Battery reverse protection

- Meters:

  Input Voltmeter with selector switches

  Input Ammeter with selector switches

  Output Voltmeter with selector switches

  Output Ammeter with selector switches

- Indications:

  R,Y,B

  Output ON

  Charger ON float

  Charger ON boost

- Alarms with indication

  Input supply fails

  Charger fails

  Rectifier fuse fails

  Battery under voltage

  Output low

  DC earth

- Ambient Temperature: 0–40^{0}C
- Enclosure: IP-20, IP-45
- Cooling: Natural air/force cooled
- Dimensions: Normally specified by the manufacturer
- Insulation level: 2 KV for 1 minute
- DC distribution board: Part of charger or separate
- Auto operation: Required/not required

8.13 Tests

Following are the routine tests as specified in the IS:10918 (For Ni-Cd) and lead-acid batteries:

1. Visual inspection
2. Checking of dimension, mass and layout
3. Marking
4. Open circuit voltage, polarity and absence of short circuiting
5. Air pressure test
6. Insulation resistance test
7. Ampere-hour test

For air pressure test, the sealing of each of the battery is checked by compressed air at a pressure equal to 150 mm H_2O. The air pressure shall be applied upto 15 second in dry and uncharged condition.

8.14 Capacity Test

For capacity test, the batteries are charged and then left for not less than 16 hours and not more than 24 hours at 27 ±5^{0}C from completion of full charge. The cell shall be discharged at the same temperature through a suitable variable resistance and at a constant current as specified by the manufacturer until the terminal voltage falls to 1.0V/Cell (Ni-Cd battery) or 1.75 to 1.85 V (Lead-acid battery). The terminal voltage of each cell, discharge current and electrolyte temperature are monitored hourly. The multiplication of rate of discharge and the time in hour will be the A-H capacity of the battery. If the temperature is different from the 27^{0}C, then the capacity shall be corrected according to the following formulas:

$C_a = C/(1+\nu(t-27))$

C_a = Corrected capacity

C = Capacity at temperature other than 27^{0}C

t = Room temperature

ν = 0.008 as per the IEC 896-2

8.15 Conclusion

The critical load like protection of electrical system, DCS system etc. is normally supplied through the DC source. Therefore a DC system, which includes battery bank, charger, distribution board etc. should be reliable and redundancy in the system should be kept. Nowadays the dilemma of design engineers is that about which type of battery bank, i.e. lead-acid or nickel-cadmium, is more reliable and give trouble-free operation.

In this chapter, the different aspects of both types of batteries are covered. The ambient temperature is the most important aspect that affects the battery life to a great extent. Therefore, it is desirable to keep the batteries at the temperature as specified by the manufacturer.

8.16 Software

Objective

The objectives of the software supplied with this book are:

- To calculate the Ampere-Hour capacity of the lead-acid battery for substation application
- To develop the specifications for nickel-cadmium and lead-acid battery

How to use it

1. Click on the "Battery" on the main menu bar.
2. Fill in the load details in the frame of "Load Details". No box should be left empty. Put "0" in the empty box.
3. Kt factor is calculated by clicking on the "Calculate" button for each cycle.
4. The Ampere-Hour capacity for different cycle is calculated, and the temperature and Ageing factors are applied.
5. Future capacity is also taken into account.
6. The battery bank nearest to the calculated Ampere-Hour capacity is chosen.

Example 8.1

The first step is to fix a load profile. We take the following 2 hours load profile

- Four circuit breakers tripped simultaneously. Each circuit breaker takes 6 amp, so the total momentary load is 24 amp.
- Emergency light load 15 amp, continuous (2 hours).
- Emergency motor starts up and running (20 amp starting and 10 amp running) (30 minutes).
- Closing of four circuit breakers, one at a time. Each breaker requires 15 amp momentary load.

Load and Duration

Load	Current (Ampere)	Duration (Minute)
Tripping of four circuit breakers simultaneously + emergency light + starting of emergency motor	24 + 15 +20 =59	01
Emergency light + running of emergency motor	10 + 15 = 25	29
Emergency light	15	89
Emergency light + closing of one circuit breaker	15 + 15 =30	01

Table 8.4 shows the ampere-hour capacity calculation of a Lead-Acid battery bank for a cyclic load.

Table 8.4

Period (1)	Current (Amp) (2)	Change in load (Amp) (3)	Duration (Min.) (4)	Time to end of section (Min.) (5)	Capacity Factor Kt (6)	A-H (+ve) (3)x(6) (7)	A-H (-ve) (3)x(6) (6)
1	59	59 - 0 = 59	01	120	3.01	177.59	
2	25	25 - 59 = -34	29	119	3.01		102.34
3	15	15 - 25 = -10	89	90	2.53		25.3
4	30	30 - 15 = 15	01	01	0.45	6.75	—
					Sub total	184.34	127.64
					Net capacity required= 184.34 – 127.64 = 56.7 Ampere-Hour		

Temperature correction factor 0.91
Aging factor 0.8
Capacity after the correction factor = 56.7 x 0.91/0.8 = 64.49
Margin = 20%
Capacity = 77.39 A-H
Nearest capacity will be of 100 amp hour.

Refer to Fig. 8.1, which shows the computer screen for the same calculation.

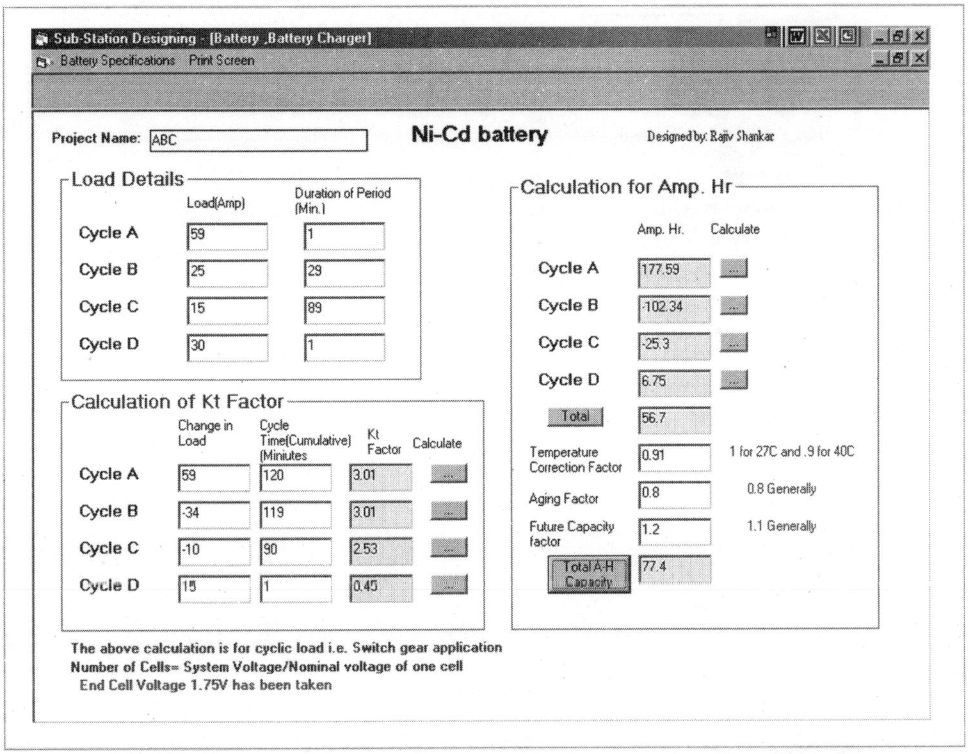

Figure 8.1: Screen for calculation of Ampere-Hour capacity of lead-acid battery

Specifications

- Click on the "Specification" menu.
- Sub-menu of "Ni-CD" and "lead-acid" opens up.
- By clicking on the respective menu the respective screen opens up. (Fig. 8.2)
- Press ADD button and fill in the boxes as per the required specification. Press UPDATE button.
- Click "Generate specification" button.

  Fig. 8.3 shows the computer screen for generation of specification in a tabular form.

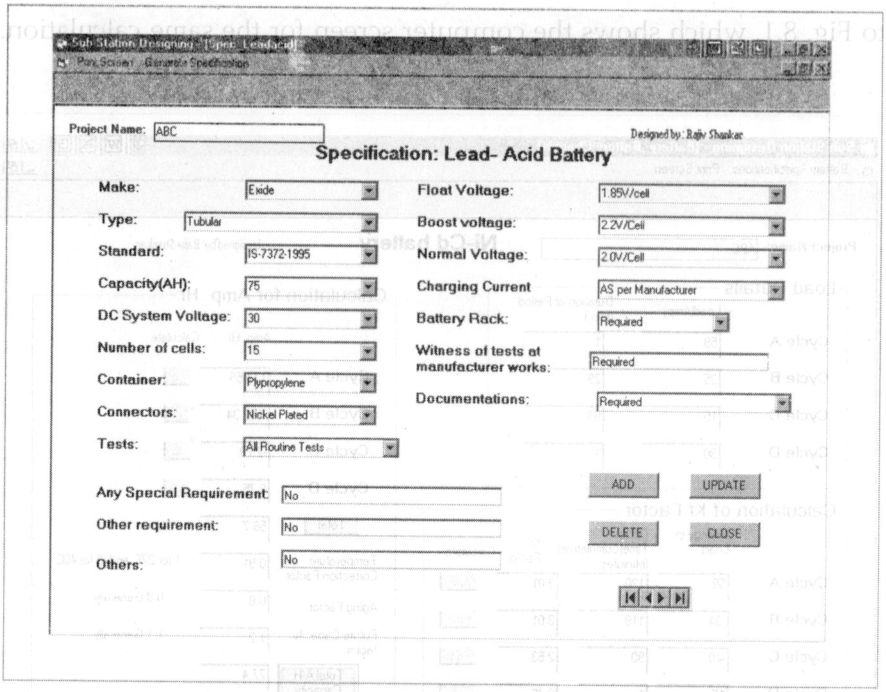

Figure 8.2: Screen for development of battery specifications

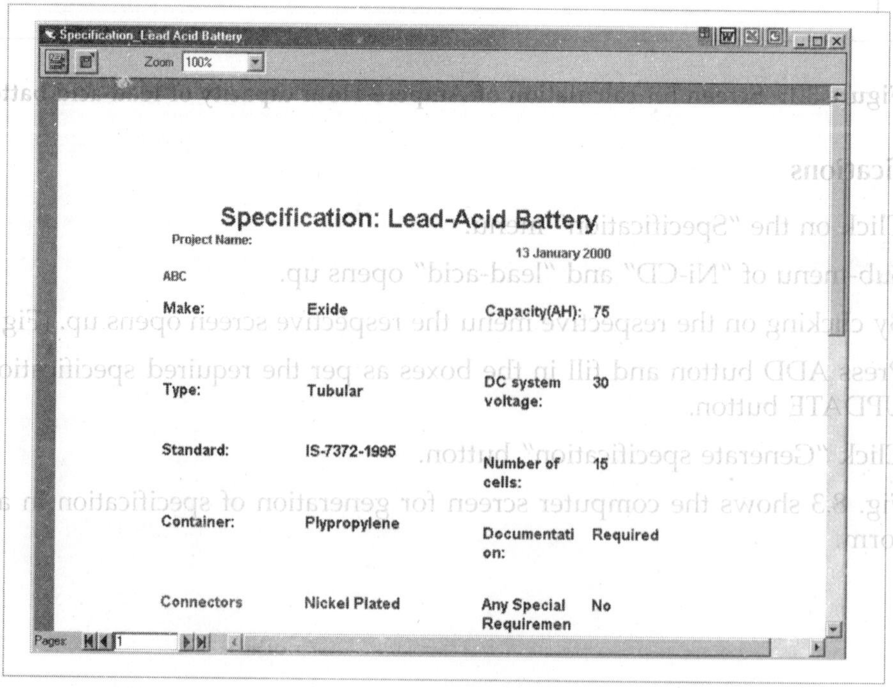

Figure 8.3: Screen for generation of specification in a tabular form

9 Diesel Generating Set

9.0 Introduction

The DG sets can have the following applications:

- Emergency power supply.
- To cater to peak load.
- Remote area where normal grid power supply is not available.

It is a most reliable source of power supply, and power is made available almost instantaneously. It can run continuously for long hours without any problems. However, after a certain period, it is required to stop the DG set to see the lube oil level.

9.1 Selection of DG Set

The following things are considered before selecting a DG set:

- Type of load
- Duty cycle
- Type of fuel
- Location of installation

9.1.1 Non-linear load

Non-inear loads are those loads which do not have constant impedance to sinusoidal varying voltage. This may be due to the inherent characteristic of load or due to the switching effect in the circuits like rectifiers, inverters etc. The current wave form of non-linear load is not sinusoidal even though the sinusoidal voltage is applied. Therefore, the internal reactance drop in the generator will be non-sinusoidal and the terminal voltage, which is the generated voltage minus the non-sinusoidal voltage drop will also be in non sinusoidal. This will adversely affect the other equipments connected with the generator. In general, the non linear load should not be more than 40% of the total load.

The effect of non-linear load can be reduced by selecting the oversized generator, since as the size i.e. KVA rating of the generator increases, the internal reactance reduces and the non-linear voltage drop reduces.

9.1.2 Capacitive load

In many installations, capacitor banks are provided to compensate the power factor and it is a practice to install capacitor bank at the LT end of the transformer for bulk compensation instead of providing capacitor for individual load. This practice sometimes leads to overcompensation and current becomes leading. The field current increases if the lagging power factor reduces, while it reduces in case of leading power factor. If the leading power factor reduces sufficiently, a condition may reach in which the field requirement may become zero. At this stage, the AVR (Automatic Voltage Regulator) sensing the generator output voltage would lose control and stability of generator will be lost. Any further reduction in leading power factor would result in over excitation and voltage will rise. To prevent such a situation, the power factor correction bank should be disconnected when supplying power supply through the generator.

9.1.3 Unbalance load

Unbalance loading causes additional heating of the generator and unbalance output voltage. It is desirable to limit the amount of unbalance to less than 10–15% of rated current. For large machines a limit is placed in neutral current, which is 50% of the line current. In closed loop voltage regulation, the majority sense only one of the three line-to-line voltage and the voltage of uncontrolled line can deviate excessively. Three-phase sensing with compounding transformer type static excitation are preferred for large capacity of generators.

In parallel operating generators to prevent circulating harmonic currents, it is customary to earth neutral of only one generator.

9.1.4 Motor load

The generator must be capable to take the starting load of the motor of the largest rating while the other load connected with the generator remains unaffected. Normally, if the motor is directly on-line starting type, it may take six to seven times of the rated current during starting for 2 to 3 seconds and a voltage drop across the generator terminal will take place. Care should be taken that the voltage drop should not reach to such an extent that the connected load gets affected. When the voltage drops beyond a certain value, generally at 65% of the normal voltage, the holding coil of the contactor will not have sufficient strength to keep the contactors in close position and the motor will trip. The magnitude of the voltage drop at generator terminals can be calculated.

9.2 Assessing the Load

The main factors which determine the choice of the generating set are:

- Maximum continuous load.
- Maximum KW which can be applied in a single step.
- Maximum value of applied KVA in a single step relative to a given transient voltage dip.

The other factors in selection of DG sets are duty cycle, i.e. continuous running, intermittent running, stand-by operation.

9.3 Location for Installation of DG Set

The ambient temperature, height from sea level etc. should also be taken care of. If the value is more than the designed values of the machine, then the de-rating factor as per the manufacturer recommendation should be applied.

9.4 Specification

The specifications of diesel engine vary from manufacturer to manufacturer. The type of application for which the D. G. Set is required should be specified, for example D. G. Set of intermittent duty is required for feeding of emergency power supply during failure of normal power supply. Similarly, for base load application the DG set should be of continuous rating.

DG sets are designed considering certain ambient conditions like ambient temperature, height from sea level and relative humidity in the atmosphere. If the location where the DG set is going to be installed has different ambient temperature, height etc. from the designed conditions, then it is required that the set should be suitably de-rated so that it can perform well under those conditions which are different from its designed values. It is important to specify the ambient temperature, height from sea level in the specifications.

The DG set has the following internal system:

- Starting system
- Engine speed governing system

- Fuel system
- Lube oil system
- Cooling system
- Air intake and exhaust system

The engine can be started either through battery or compressed air as the case may be.

9.5 Conclusion

DG sets are generally used for supplying emergency power requirement in any industrial, commercial or domestic establishment. AMF (Auto Mains Failure) panel is provided along with (Auto mains failure) the DG sets to start the engine automatically on failure of the mains power supply. The emergency load is connected with the DG sets through the auto changeover switch. The changeover from normal circuit breaker to emergency circuit breaker takes place automatically after the alternator builds up the voltage.

The most important aspect is the selection of a DG set of appropriate rating, and this task is quite difficult since the loads in commercial and domestic complexes vary widely. If the DG set of higher capacity is chosen, it will result in operation of DG set at low load. The operation of DG set at low load is quite harmful. It will result in improper burning of fuel resulting in carbonization of the cylinders and others parts of the set. The efficiency at low load will be low and the operation of the DG set will not be economical. Similarly, operation of DG set at over-capacity will again result in uneconomical operation and the overall life of the set will be affected. The optimum loading is between 50–75% of the capacity. Usually, the manufacturer of the DG sets is in a position of supplying the fuel consumption curves at different load.

It is always desirable to have two DG sets of equal capacity and system should be such that when the load is less only one DG set will start, and as the load increases beyond a set point the second set also starts, both the sets are synchronized and the sharing of load between the two DG sets will take place.

The other important aspect in deciding the capacity of the DG set is the sequence of loading the DG set and the method of starting the motors connected with it.

9.6 Software

Objective

- To calculate the capacity of the DG set.
- To calculate the voltage drop across the DG set terminals after putting a large load, for example starting of an induction motor through DOL starter.
- To develop specifications of the diesel engine and the alternator.

How to use it

1. Click on the "DG set" on the main menu bar. (Fig. 9.1)
2. Fill in the required load details in the respective boxes and calculate the total connected load (KVA)
3. The chosen rating should depend on the maximum demand and not on the transient and connected load.

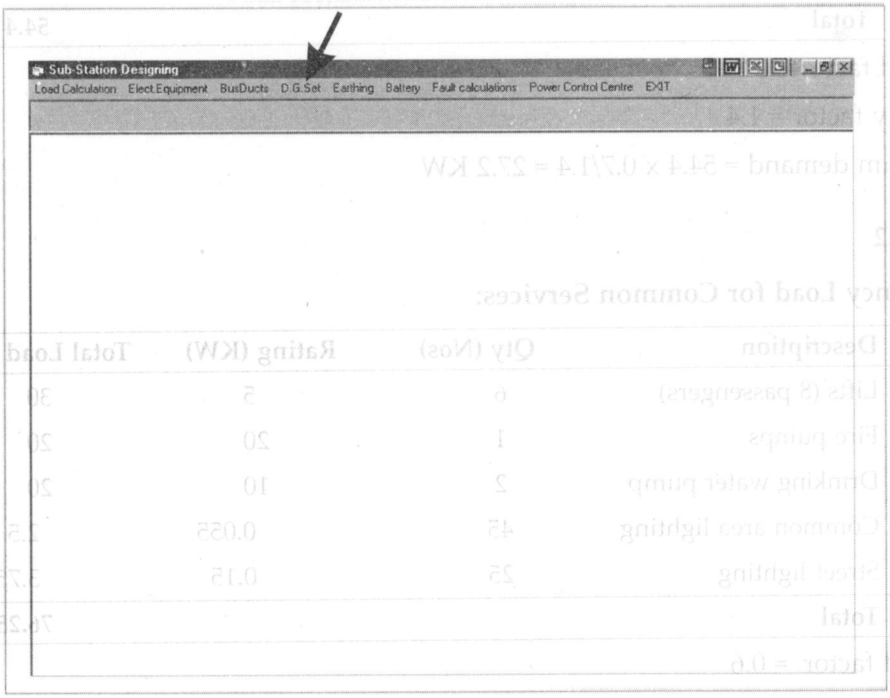

Figure 9.1: Main screen

4. The voltage drop during application of transient load can also be calculated to ascertain that the voltage drop is within the limit. If the voltage drop is more than the specified limit then the higher rating of set should be selected.

Example 9.1

Emergency load tables from Chapter 1 are reproduced below. (Refer to Table 9.1 and 9.2). The objective is to calculate the capacity of the DG set to feed the loads listed below.

Table 9.1

Sl. No.	Description	No. of dwelling units	Load/Flat(KW)	Total Load(KW)
1.	**BLOCK-1**			
	Type-1	14	0.5	7.0
	Type-2	14	0.6	8.4
2	**BLOCK-2**			
	Type-3	18	0.7	12.6
	Type-4	18	0.8	14.4
3.	**BLOCK-3**			
	Type-5	12	1.0	12.0
	Total			**54.4**

Demand factor = 0.7

Diversity factor = 1.4

Maximum demand = 54.4 x 0.7/1.4 = 27.2 KW

Table 9.2

Emergency Load for Common Services:

Sl. No.	Description	Qty (Nos)	Rating (KW)	Total Load (KW)
1.	Lifts (8 passengers)	6	5	30
2.	Fire pumps	1	20	20
3.	Drinking water pump	2	10	20
4.	Common area lighting	45	0.055	2.5
5.	Street lighting	25	0.15	3.75
	Total			**76.25**

Demand factor = 0.6

Diversity factor = 1.5

Maximum demand = 76.25 x 0.6/1.5 = 30.5 KW

Total Emergency Load = 27.2 KW + 30.5 KW = 57.7 KW

Considering 30 per cent future increase in demand, total maximum demand will 1.3 x 57.7 = 75.01 KW. The DG set, should be chosen in such a way that it is capable of supplying 75.01 KW maximum load.

The nearest capacity will be 100 KW or 125 KVA at 0.8 power factor.

Now, the following question arises. What will be the voltage dip if the motor of the largest capacity is started while the generator is running with its base load?

The base load is lighting load, lift load and emergency load of houses are as given below:

Emergency housing load: 27.2 KW

Lifts: 12KW

Area lighting: 1.0 KW

Street lighting: 1.5 KW

Total: 41.7 KW = 52.15 KVA (assuming p.f.= 0.8)

i.e base load = 72.55 amp

Let's take starting current of 10 KW motor = 90 amp (DOL starting)

Total load = 90 + 72.55 = 162.55 amp

Voltage dip = Xt/(Xt+ C)

Xt = per unit transient reactance

C = generator rating KVA (or rated current)/impact load KVA (or step load in ampere)

Per unit unsaturated transient reactance of 125 KVA alternator = 0.25 (to be obtained from the manufacturer)

C = 174/162.55 = 1.07

Voltage dip = 0.25/(0.25 + 1.07)

= 0.25/1.32 = 0.1893 per unit

= 18.93%

Since the load is housing load and there is no sensitive load, this voltage dip is acceptable. This is an approximate calculation. Manufacturer of the DG set supplies the curves for calculation of the voltage dip and more accurate results can be obtained from the curves.

Now the question arises as to what will happen if the dip is more or some sensitive load like solenoid valves, electronic controllers etc, which are more common in a process industry, are connected with the generator and the voltage dip more than the specified limit takes place. In that case we have two options.

- To increase the capacity of the DG set.
- To supply the sensitive load from other source like UPS system.

Refer to Fig. 9.2, which shows the computer screen where calculations of Example 9.1 are done through the software.

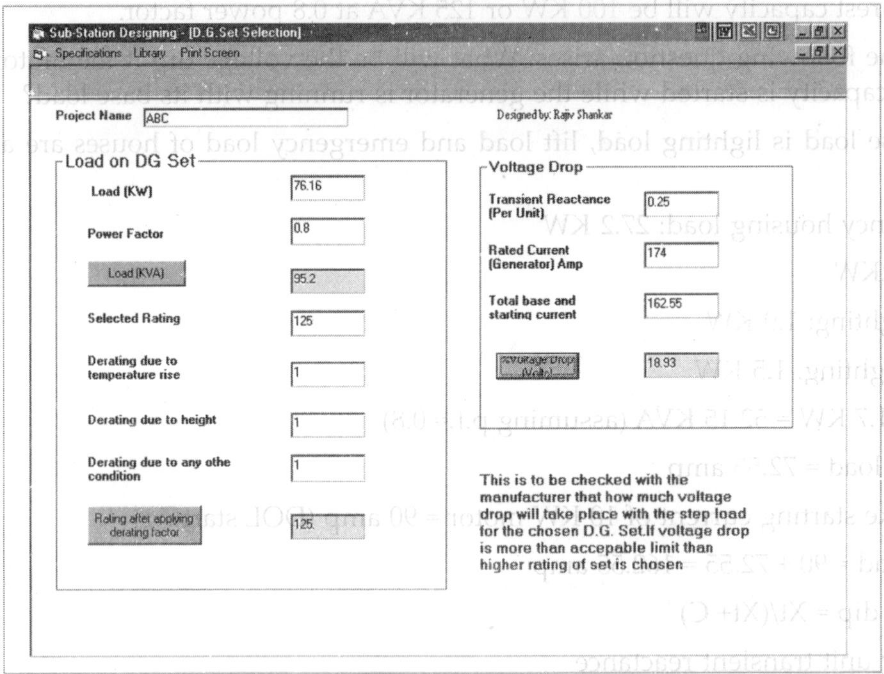

Figure 9.2: Screen for calculation of capacity of DG Set and percentage voltage drop

Specifications

Specifications of diesel engine and alternator and control panel can also be developed by clicking on the sub-menus of "Alternator" and "Engine" respectively. (Refer to Fig. 9.3 and 9.4). The procedure is same as mentioned in the earlier chapter.

Library

The parameters of DG set can be stored, retrieved, deleted by the user.

To perform the above task the following steps are to be followed:

- Click on the "Library Entry" on the menu bar.
- Click ADD button and fill in the required data in the specified boxes.
- Click the "Update" button and close the screen by clicking the close button.

- To retrieve the data click on the "Library" on the main menu.
- A screen opens up containing the required data. (Refer to Fig. 9.5 and 9.6).

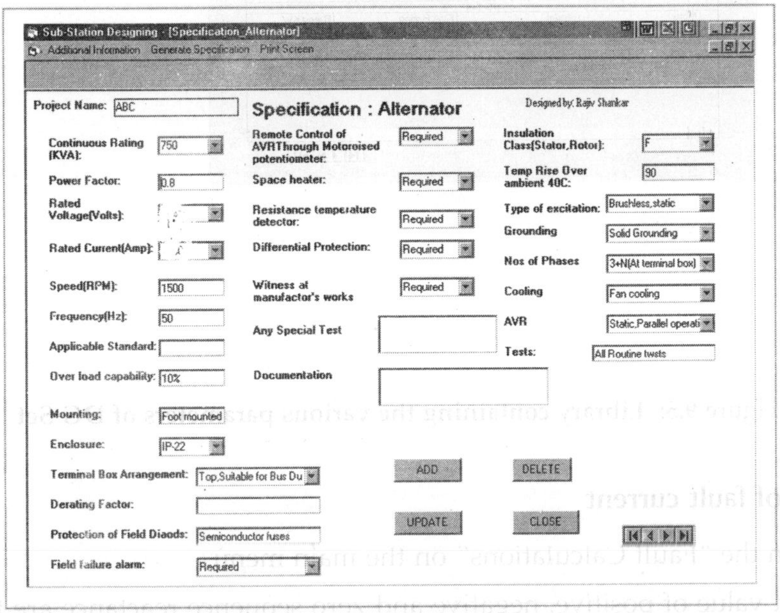

Figure 9.3: Screen for development of specification of alternator

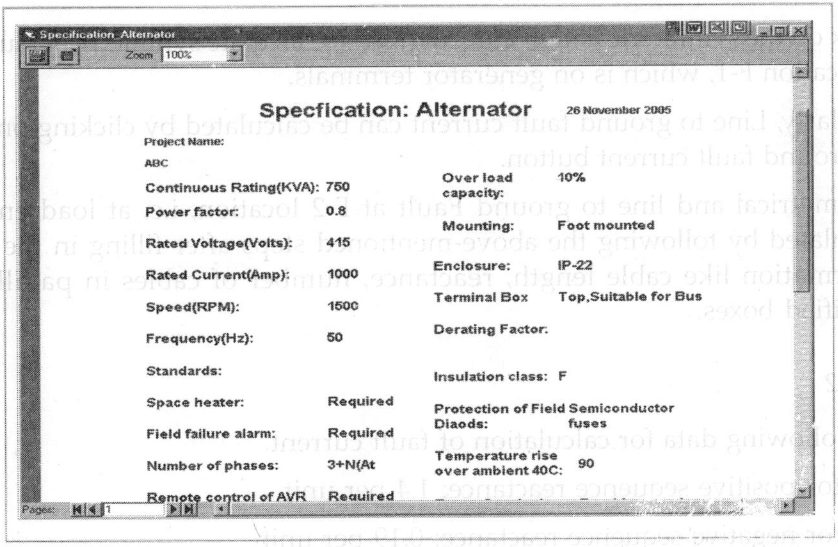

Figure 9.4: Screen for generation of specification in a tabular form

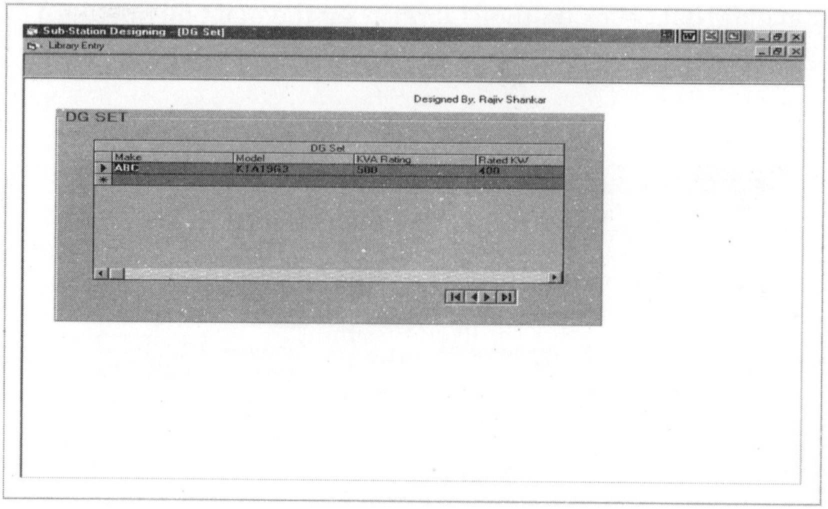

Figure 9.5: Library containing the various parameters of DG Set

Calculation of fault current

- Click on the "Fault Calculations" on the main menu.
- Per unit value of positive, negative and zero sequence reactance are to be filled in the specified boxes.
- Click on the fault F-1 Tab.
- Fill in the Generator KVA value.
- Click on the symmetrical fault current (KA) to calculate symmetrical fault current at location F-1, which is on generator terminals.
- Similarly, Line to ground fault current can be calculated by clicking on the Line to Ground fault current button.
- Symmetrical and line to ground Fault at F-2 location, i.e. at load end can be calculated by following the above-mentioned steps after filling in the required information like cable length, reactance, number of cables in parallel in the specified boxes.

Example 9.2

Take the following data for calculation of fault current.

1. Generator positive sequence reactance: 1.4 per unit
2. Generator negative sequence reactance: 0.19 per unit
3. Generator zero sequence reactance: 0.106 per unit
4. Generator rating: 750 KVA

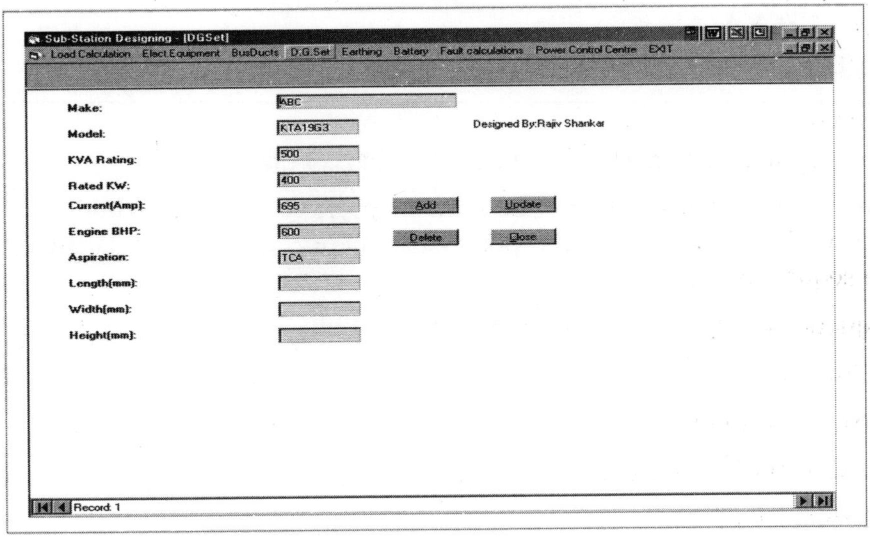

Figure 9.6: Library for starting of DG Set

5. Length of cable (Generator PCC to Load PCC): 300 metres
6. Number of cables in parallel: 04
7. Cable reactance: 0.05 ohm/km

Fault at F-1

X1 (Positive sequence reactance) in ohm = 1000 x per unit value x KV^2/KVA

$= 1000 \times 1.4 \times 0.415^2/750$

= 0.32 ohm

X2 ((Negative sequence reactance) in ohm

$= 1000 \times 0.19 \times 0.415^2/750$

= 0.043 ohm

X0 (Zero sequence reactance) in ohm

$= 1000 \times 0.106 \times 0.415^2/750$

= 0.024 ohm

Symmetrical fault current:

E phase- N/X1

= 240/0.32 = 746.53 = 0.75 KA

Line to ground fault current:

= 3 x E phase-N/(X1 + X2 + x3)

= 3 x 240/ (0.32 + 0.024 + 0.043)

= 1860.46 = 1.86 KA

Cable Reactance = 300 x 0.05/(1000 x 4)

= 0.00375 ohm

X1C (Positive sequence reactance of cable) = 0.00375 ohm

Negative sequence reactance will be same as positive sequence while zero sequence reactance will be 3 to 5 times the positive sequence, assuming it is three times the positive sequence reactance.

Zero sequence reactance = 3 x 0.00375 = 0.01125 ohm

Fault at F-2:

Symmetrical fault current

= 240/ (0.32 + 0.00375) = 741.3 A = 0.74 KA

Line to ground fault current

=3 x 240/ ((0.32 + 0.00375) + (0.043 + 0.00375) + (0.024 + 0.01125))

= 1781.076 A = 1.78 KA

All the above calculations are also done by using the software as shown in Fig 9.7 and 9.8.

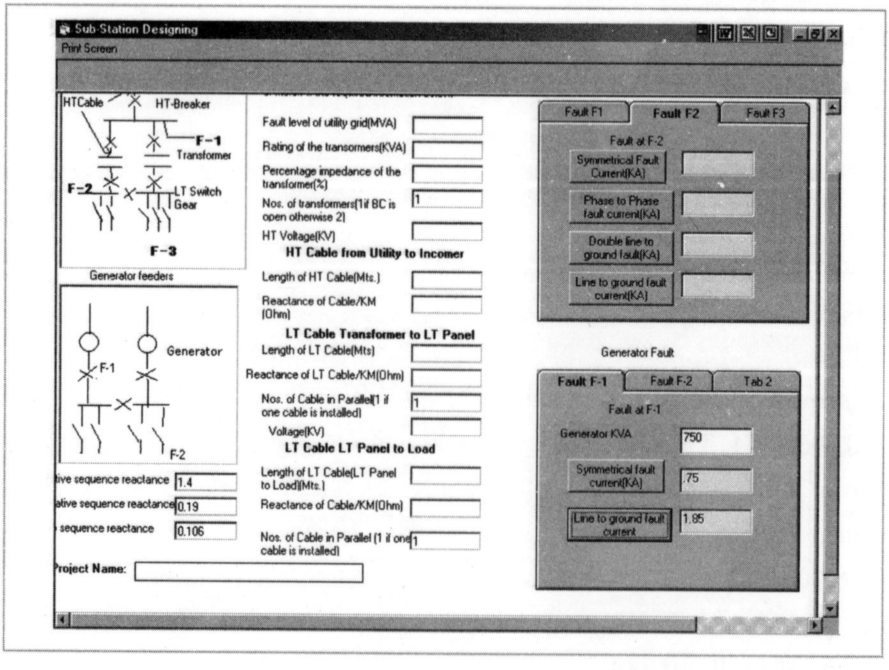

Figure 9.7: Screen for calculation of fault current at generator terminals

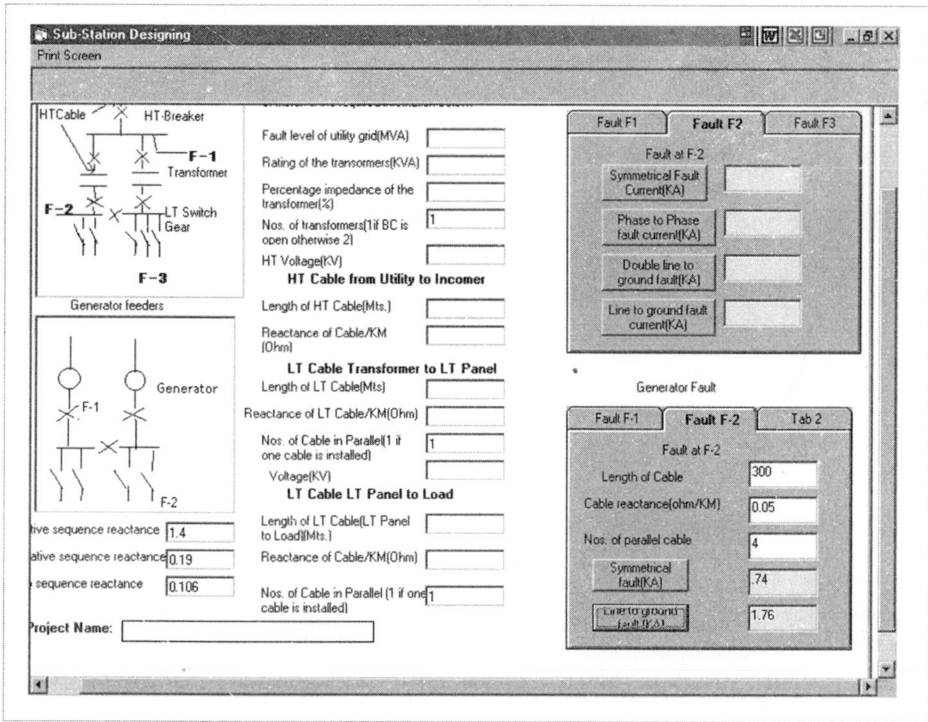

Figure 9.8: Screen for calculation of fault current at load end

References and Bibliography

1. *Electrical Systems Analysis and Design for Industrial Plants* (1985): Irwin lazar (McGraw-Hill)
2. *Handbook of Electrical Engineering, For Practitioners in Oil, Gas and Petrochemical Industry.* (2003) Alan L. Sheldrake (John Wiley & Sons)
3. *Diesel Generator Handbook* (1992) LLJ Mahon (Butterworth–Heinemann)
4. *Network Protection & Automation Guide,* ALSTOM (2002) T&D Energy Automation & Information, France.
5. *Modern Power Station Practice,* Volume 4 Electrical (Generator & Electrical Plant) Pregaman Press
6. *J&P Switchgear Hand Book* (11th Edition) R.T. Lythall, (Buttorworth & Company ltd.)
7. *Power Station Protection Volume 1 & 2,* The Electricity Council, UK (Peter Peregriums, UK)
8. *Electric Power Distribution Systems* A.S. Pabla (Tata Mc Grawhill)
9. *J&P Transformer Book* (11th Edition) A.C. Franklin, D.P. Franklin (Buttorworth & Company)
10. *Elements of Electrical Power Station Design* (1985) Prof M.V. Deshpande (A. H. Wheeler & Co.,

Indian Standards

Cables

IS 694:1990 PVC insulated cable for working voltage up to and including 1100 volts-Specifications.

IS 1255-1983 Selection of cables.

IS 7098 (Part-1)-1988 Specification for cross linked polyethylene insulated PVC sheathed cable.

Transformer

IS 6600-1972 Guide for loading of oil-immersed transformers.

IS 10028 (Part-I, Part-II, Part-III) –1981 Code of practice for selection, installation and maintenance of transformers.

IS 10561- 1996 Application guide for power transformer.

IS 11171-1996 Specification for dry-type power transformer.

IS 2026 (Part-I) 1984 Specification of power transformer (General)

IS 2026 (Part-II) 1984 Specification of power transformer (Temperature rise)

IS 2026 (Part-III) 1984 Specification of power transformer (insulation level and dielectric test)

IS 2026 (Part-IV) 1984 Specification of power transformer (terminal markings, tappings and connections)

IS 335: 1993 New insulating oil-specification

IS 1866: 2000 Code of practice for electrical maintenance and supervision of mineral insulating oil in equipments

IS 9434: 1992 Guide for sampling and analysis of free and dissolved gases and oil from oil-filled electrical equipment.

IS 10593: 1995 Method of evaluating the analysis of gases in oil-filled electrical equipment in service

Batteries

IS 1651: 1991 Stationary cells and batteries, lead-acid type (with tubular positive plates) – Specification.

IS 1652: 1991 Stationary cells and batteries, lead-acid type (with plante positive plates) – Specification.

IS 6304: 1992 Stationary batteries lead-acid type pasted positive plates – Specification.

IS 10918: 1984 Specification for vented type nickel-cadmium batteries.

Switchgear

IS 10118 (Part I): 1982 Code of practice for selection, installation and maintenance of switchgear. (General)

IS 12021-1987: Specification for control transformers for switchgear and control-gear for voltages not exceeding 1000 V ac.

IS 12729–1988: Switchgear and control gear for voltages exceeding 1000 V – General requirements.

IS 2516 (Part IV/Sec.2): 1980 Specification for circuit breakers (type tests and routine tests).

IS 13947 (Part-2): 1993 – Specification for low-voltage switchgear and control-gear. (Circuit breakers)

IS 13947 (Part-3): 1993 – Specification for low-voltage switchgear and control-gear. (Switches, disconnectors, Switch disconnectors and fuse combination unit)

IS 13947 (Part5/Sec 1): 1993 – Specification for low-voltage switchgear and control-gear (Electromechanical control circuit devices)

INDEX

Absorbed Glass Mat (AGM) 167
Accessories 34, 71–2, 177
Acidity test 38
Active material 168–9, 175
Active power component 138
Ageing factor 172–3, 181
Air circuit breaker 67, 70–2, 75, 88, 92
Air gap 138
Air insulated 126
Ambient condition 13, 105, 125–6, 187
Ambient temperature 1, 12–13, 32, 34, 47, 68, 77, 109, 112, 126–7, 131, 170, 179, 181, 187
Ampere-hour capacity 181–3
Annealed high conductivity 106
Annual capitalized cost 48
Annual running cost 45–6, 51
Annunciation 57, 79, 82
Apparent power 138
Arc quenching medium 61, 76, 81
Area classification 13
Assessment of load 6
Auto change-over 59, 83–4, 88, 92, 188
Auxiliary losses 31, 42–3
Auxiliary relay 78, 80
Base KVA 85–6, 99, 100, 118, 120
Base load 89–90, 187, 191
Battery bank 167, 169–71, 173, 175, 177, 179–83
Bending radius 109
Bimetallic effect 127
Bimetallic strip/relay 68
Bimetallic washers 106
Boost voltage 177–8
Breaking current 64–7, 73, 82
Bridge rectifier 178
Bus-bar 10, 57–61, 78–9, 82–3, 85–6, 92, 96, 126–9, 131–2, 139
Bus-duct 2, 12, 57, 77, 95, 125–32
Cable 2, 10, 12, 36, 43, 47, 57, 68, 77, 82, 84–5, 89, 91, 93, 95, 98, 99, 105–21, 123, 125, 131, 140–1, 144–5, 156–8, 160, 194–6, 201
Cables 109
Capacitor bank 11, 17–18, 137–47, 186
Cell voltage 169, 172–4, 176
Charge cycle 168

Charger 12, 167, 170, 172, 174–5, 177–80
Circuit-breaker 10, 36, 40, 57, 59, 61–7, 67–75, 79–88, 85–6, 88, 92–6, 101–2, 118, 127, 151–2, 155, 157, 172, 181–2, 188, 202
Coordination (Relay) 65–6, 69, 71, 76, 78, 81, 85, 91, 94–5, 97, 156
Critical bus 79
Cut off 36, 76, 169, 174
Deep cycle battery 167
Deep discharge 176
Dehydration 36–38
Demand factor 2, 5, 16–18, 23, 26, 44–5, 115, 190
De-rating factor 110, 112–14, 126, 129, 131, 187
DG set 1, 18, 84, 95, 167, 185, 187–92, 194–5
Diesel engine set 187, 189, 192
Diode dropper 178
Discharge cycle 167, 174
Dissolved gases 33, 38, 202
Duplicate bus bar 60–1
Duty cycle 2, 169, 185, 187
Earth electrode 157–8, 162
Earth leakage 70–1, 158
Earth pit 156, 161
Earth/ground fault 72, 76, 78, 80, 82, 89, 99, 100, 107, 149, 150–58, 160–1, 164, 194–6
Electrical equipment 11–13, 48, 102, 113, 126, 137–38, 145, 149, 202
Electrolyte 167, 169, 171, 174–7, 180
Emergency load 1–5, 17-18, 22, 83, 87, 167, 188, 190–1
Equipment 1–2, 5, 10–13, 13, 22, 26, 30–31, 36, 43, 48, 57, 77–8, 95, 97, 102, 105–6, 113, 126, 137–8, 145, 149–50, 153, 156, 161, 172, 186, 202
Explosion proof equipments 13
Fault current 40, 62, 64, 69, 71, 76, 79, 80, 84–5, 89, 93, 97–8, 100–2, 107, 127, 128, 131, 149, 151, 153–8, 160–1, 164, 194–7
Fault level 1, 64, 80, 82, 118, 131
Fire resistant 10
Fire separator 36
Float charging 173, 178
Fuse 57, 67, 69, 71, 75, 92, 94, 127, 151, 178, 179, 202
Grounding 10, 92, 126, 129, 130, 149–51, 153–61, 164
Grounding conductor 156, 160, 164
Group compensation 141
Harmonic current 42, 156, 186
Heat run test 92
High resistance fault 150
High transient 150
High voltage 9–10, 11, 29, 64, 91, 92, 105, 109, 150, 156, 174
Impulse voltage 81
Indication 57, 79, 179
Induction level 105
Industrial load 3, 5, 10, 23, 26
Inflammable gases/environment 64, 109
Inrush current 40–1, 48
Insulation level 67, 179, 201
Interlocking 10, 57, 92
Ions 63, 169, 175–6
Joints (busbar) 10, 78, 82, 128–30
K-factor 172
Lagging (power factor) 138, 186
Laying (cable) 12, 109–10, 112
Layout (substation) 11–12, 180
Lead oxide 167–8
Lead sulfate 168, 177
Lead-acid battery 167–9, 173, 175–6, 180–3
Leading (power factor) 186
Losses (transformer) 4, 9, 31, 34, 42–5, 47–8, 50–51, 54, 109, 111, 113, 115–17, 137, 139, 144–7
Making current 67, 75, 82
Material 30, 33–4, 38–9, 47, 68, 106–8, 110, 125–7, 130–1, 160, 164, 168–9, 175
Measuring device 93, 96, 81–2
Medium (arc quenching) 61, 76, 81
Medium voltage 9–10, 151, 153, 157
Memory effect 174–5
Metal clad 57
Metal ions 63
Microprocessor base release/relay 70, 72, 75, 80–2, 96
Microprocessor based multi parameter measuring device 81–2, 93, 96
Miniature circuit breaker 67–71, 178
Moulded case circuit breaker 10, 57, 67, 69, 70–1, 75–6, 88, 93–5
Negative sequence reactance 98–9, 194–6

Negatively charged ions 63
Nickel-cadmium battery 168, 173–5
Nominal voltage 169
Non linear load 185–6
Non sinusoidal voltage drop 186
Off load tap changer 30, 32, 43, 47
Oil circuit breaker 64, 80, 93
On load tap changer 29–30, 32, 43, 60, 95
Over current relay/protection 40, 78, 80, 82, 90, 150–3
Over voltage 150
Oxidation of oil 38
Pay back period 111, 139, 143–4, 146
Percentage impedance 29, 31, 40, 47–9, 54, 118–20
Phaser group test 37
Pipe electrode 159, 161–2
Plante plate cell 168, 176, 202
Plate electrode 159, 161
Polarity test 37
Protection scheme 1, 57, 78–9
Protective device 61, 67, 69, 71, 74–5, 151, 156, 158
Reactance 50, 84–6, 89, 98–100, 112, 118–20, 130, 150, 153–4, 186, 191, 194–6
Reactive power 137–38, 140
Reference ambient temperature 13, 77, 126
Regulation voltage 9, 29, 31, 48, 137, 186
Releases 39, 71–2, 76
Residential load 11
Resistance 37, 38, 50, 81, 107, 109, 111, 112, 115, 127–30, 144, 150, 151, 153–9, 161–3, 165, 171, 180
Resistive grounding 150
Safety requirements 2, 36, 79
Sectionalized bus-bar 59, 61
Selection 12, 29, 31, 35, 43, 48, 51–2, 62, 69, 71–2, 81, 96–7, 105, 108, 111, 113–15, 125, 145, 169, 185, 187–8, 201, 202
Short circuit rating 31, 62, 93, 110, 113, 125, 130
Single bus-bar 58
Sizing calculation of bus-bar 125
Specifications 1, 13, 29, 34, 43–4, 46, 48, 52–3, 57, 67–68, 70–3, 75, 77, 81, 96–7, 102, 113, 120–22, 125–6, 130, 132, 145, 147, 161, 165, 176, 178, 181, 183–4, 187, 189, 192, 201
Substation space (indoor/outdoor) 1, 11–12, 30, 32, 35–6, 43, 46, 62, 79, 81, 129
Sulfate ions 175–6
Switchgear 11, 57, 80, 95–6, 118, 120, 167, 169, 172, 176, 178, 202
System 2–6, 9–11, 15, 19–23, 29, 36, 43, 57–62, 64, 67–9, 71–2, 78–85, 88–9, 92–8, 102, 105, 108, 110–11, 113, 118, 125–8, 131, 137, 139, 141–51, 160–1, 169, 173, 180, 187–8, 192
Temperature rise 2, 11, 13, 34, 47, 78, 107, 110, 126–31, 156, 160, 170, 201
Testing of transformer oil 33, 37-8
Transformer 9–10, 12, 18–19, 21, 29, 30–40, 42–54, 57, 78–86, 89–90, 92, 95, 98–9, 105, 118, 120, 125, 137–8, 140–1, 145, [illegible] 186, 201
Transmission losses 4, 9
Trends (future trends in switchgear) 57, 92, 96
Tubular cell 168, 170, 176, 202
Type 1 & 2 coordination 94–5
Ultimate short circuit capacity 73
Ungrounded system 149–50
Utilization category 69, 71, 73–5
Vacuum circuit breaker 63–4, 80, 93
Ventilation 11, 35, 170
Voltage 1, 4, 9–12, 20, 21, 29–31, 34, 37, 40–3, 47–51, 54, 62, 64–72, 75, 78, 80–82, 87, 88, 91–94, 105, 108–9, 111–16, 118, 120–21, 125, 127, 130–31, 137, 149–51, 153-8, 169, 171–80, 185–92
Voltage drop 12, 40, 48–51, 109, 111–16, 118, 131, 186–7, 189, 190, 192
Voltage regulation 9, 29, 31, 48, 137, 186
Wet battery 170
Zero sequence reactance 98–9, 194–6
Zero sequence resistance 153
Zero sequence transformer 153
Zig-zag transformer 150, 154–5
Zone protection system 93–4

Installation of the software

1. The CD-ROM contains a folder "SSD".
2. Copy the "SSD" folder from the CD to C:\WINDOWS.
3. Open the "SSD" folder.
4. Click on "Edit" on the main menu bar and click on "Select All".
5. Click on "File" on the main menu bar and then click on "Properties".
6. Go to "Attributes" on the "Property" dialogue box.
7. Deselect "Read-only" and select "Archive" and click on "Apply" and then click on OK box. The "Property" dialogue box will close.
8. Open the "Package" folder and double click on "Setup.exe" icon.
9. Sub-station Designing screen will open up. Click on OK button.
10. Click on the installation button of "Station-designing software".
11. After a little while a message box with the message "Sub-station designing set up was completed successfully" will appear.
12. Close the folder.
13. Go to the program and click on "Sub-station designing" to run the program.
14. It is necessary to copy the folder in C:\WINDOWS, otherwise the functions like "specifications" and "library" will not work.